PERSPECTIVES IN ETHOLOGY

Volume 4

Advantages of Diversity

CONTRIBUTORS

Peter G. Caryl
Department of Psychology
University of Edinburgh
Edinburgh, Scotland, U.K.

Hugh Drummond
Department of Psychology
University of Tennessee
Knoxville, Tennessee 37916

Dennis Robert Rasmussen
Sub-Department of Animal Behavior
University of Cambridge
Madingley, Cambridge CB3 8AA, U.K.

Steven P. R. Rose
Brain Research Group
Department of Biology
The Open University
Milton Keynes MK7 6AA, U.K.

Wolfgang M. Schleidt
Department of Zoology
University of Maryland
College Park, Maryland 20742

P. J. B. Slater
Ethology and Neurophysiology Group
School of Biology
University of Sussex
Brighton BN1 9QG, U.K.

J. E. R. Staddon
Departments of Psychology and Zoology
Duke University
Durham, North Carolina 27706

Nicholas S. Thompson
Department of Psychology
Clark University
Worcester, Massachusetts 01610

Steven Vogel
Department of Zoology
Duke University
Durham, North Carolina 27706

R. Haven Wiley
Department of Zoology
University of North Carolina
Chapel Hill, North Carolina 27514

Kenneth R. Wing
School of Public Health
School of Law
University of North Carolina
Chapel Hill, North Carolina 27514

PERSPECTIVES IN ETHOLOGY

Volume 4

Advantages of Diversity

Edited by

P. P. G. Bateson

Sub-Department of Animal Behaviour
University of Cambridge
Cambridge, England

and

Peter H. Klopfer

Department of Zoology
Duke University
Durham, North Carolina

PLENUM PRESS • NEW YORK AND LONDON

Library of Congress Cataloging in Publication Data

Bateson, Paul Patrick Gordon, 1938-
  Perspectives in ethology.

  Vol. 3 has special subtitle: Social behavior; v. 4 has special subtitle: Advantages of diversity.
  Includes bibliographies.
  1. Animals, Habits and behavior of. I. Klopfer, Peter H., joint author. II. Title.
QL751.B188   591.5   73-79427
ISBN 978-1-4615-7577-1   ISBN 978-1-4615-7575-7 (eBook)
DOI 10.1007/978-1-4615-7575-7

© 1981 Plenum Press, New York
Softcover reprint of the hardcover 1st edition 1981
A Division of Plenum Publishing Corporation
233 Spring Street, New York, N.Y. 10013

PREFACE

One of the attractive features of the great classical ethologists was their readiness to ask different kinds of questions about behavior — and to do so without muddling the answers. Niko Tinbergen, for instance, was interested in the evolution of behavior. But he also had interests in the present-day survival value of a behavior pattern and in the mechanisms that control it from moment to moment. Broad as his interests were, he clearly separated out the problems and recognized that questions about the history, function, control, and development of behavior require distinct approaches — even though the answers to one type of question may aid in finding answers to another.

The open-minded (and clear-headed) style of ethologists like Tinbergen was based on a recognition that there are diverse ways of usefully conducting research on behavior. This consciousness has been partially submerged in recent years by new waves of narrowly focused enthusiasm. For instance, the study of the behavior of whole animals without recourse to lower levels of analysis, and the treatment of sociobiological theories as explanation for how individuals develop, has meant that the relatively fragile plants of neuroethology and behavioral ontogeny have almost disappeared under the flood.

When the floodwater starts to recede, it will doubtless leave behind it pieces of inappropriate theory stuck in the branches of the surviving disciplines. What, though, is appropriate theory? How can productive links be fostered between different subjects or between different analytical levels of a single subject? These are not easy questions to answer — particularly in the abstract. Nevertheless, we felt that the time had come to combat actively the chauvinism which can so easily become associated with a particular approach to a problem. We asked the contributors to this volume to do their bit toward breaking down the barriers that exist among the various ways in which behavior is studied. In addition, some of the contributors who were not ethologists were asked to discuss ways in which the study of behavior relates to their own disciplines.

This book starts with a contribution from Drummond on the problem of measurement. He goes a considerable distance toward resolving the tension that exists between those who believe in absolute units of behavior and those who treat their measures as merely reflecting the problem in which they are interested. Slater's plea for tolerance is addressed primarily to those who insist on always dealing with data from large samples. He argues for the benefits that come from studying individuals.

The next five contributions by Thompson, Rasmussen, Wiley, Staddon, and Schleidt relate to the enormous impact which evolutionary theory has made on ethology — particularly in the last ten years. Thompson points out that propositions about evolution — particularly in their grandest dress — are frequently circular and vacuous. Just as understanding of learning has benefitted from a focus on specific instances and from studies of the conditions that are necessary for change, so studies of survival value can move forward constructively in similar fashion. Rasmussen touches on a related issue when he wrestles with the meaning of fitness. He also emphasizes the useful points of contact between evolutionary ideas and studies of development with particular reference to work on primates. Linking approaches is also a feature of Wiley's chapter, and his concept of ontogenetic trajectories has implications for both sociobiological and developmental theory. Staddon's speculations on why humans should be so intelligent raises the important point that an understanding of how the existing machinery works may provide a useful starting point for an historical explanation. Finally, Schleidt's contribution emphasizes that if we are to understand process we must be prepared to shift our mode of thought and move back and forth from one analytical level to another. This is a point that recurs in the next group of chapters.

The articles by Rose, Vogel, and Caryl relate to three very different points of contact between ethology and other disciplines. Rose is a biochemist who has had a long-standing interest in understanding behavior in neural terms. He knows as well as anybody the problems of logic, language, and research style which can arise in moving from one level to another. Nevertheless, his perspective provides a challenge to those ethologists who wish to insulate themselves from any knowledge of neurobiology. Vogel offers a quite different challenge to ethologists who know nothing of scaling effects or of physics. He argues that such ignorance is likely to prove costly and provides some illustrations of how rewarding it can be to relate the characteristics of an animal to those of its physical environment. Caryl writes about a quite different interface — the application of mathematical game theory to the study of conflict. His concern is to show how the theory does point to analyses which would not otherwise have been done, but also how the results of such work do have implications for theory.

Once again, the importance of dialogue (in this case between the theorist and the data collector) is emphasized.

The final chapter concerns a totally different dialogue from the others, namely that between science and the law of the land. It is an issue which ethologists would do well to think about — particularly when legislation on, for instance, animal welfare is liable to impinge on their work. Wing, who is himself a lawyer, argues that laws are neither totally arbitrary nor the reflections of some highly rational process that scientists sometimes believe they are engaged in. Laws are expressions of a public consensus about the world. Ethologists have something to contribute to this set of shared beliefs, but to influence the development of the laws regulating human conduct they must understand how legal systems derive their present character.

Patrick Bateson
Peter H. Klopfer

CONTENTS

Chapter 2

INDIVIDUAL DIFFERENCES IN ANIMAL BEHAVIOR

P. J. B. Slater

Chapter 3

TOWARD A FALSIFIABLE THEORY OF EVOLUTION

Nicholas S. Thompson

Chapter 4

EVOLUTIONARY, PROXIMATE, AND FUNCTIONAL PRIMATE SOCIAL ECOLOGY

Dennis Robert Rasmussen

Chapter 8

FROM CAUSATIONS TO TRANSLATIONS: WHAT BIOCHEMISTS
CAN CONTRIBUTE TO THE STUDY OF BEHAVIOR

Steven P. R. Rose

Chapter 9

BEHAVIOR AND THE PHYSICAL WORLD OF AN ANIMAL

Steven Vogel

Chapter 10

ESCALATED FIGHTING AND THE WAR OF NERVES: GAMES
THEORY AND ANIMAL COMBAT

Peter G. Caryl

Chapter 1

THE NATURE AND DESCRIPTION OF BEHAVIOR PATTERNS[1]

Hugh Drummond[2]

Department of Psychology
University of Tennessee
Knoxville, Tennessee 37916

I. ABSTRACT

Confronted with the complex flux of phenomena that constitute the stream of behavior, the observer arrives at a verbal description by selecting information at several levels. To understand the process of selection we must come to grips with the nature of behavior patterns, which are the units of ethological science. In the analysis presented below behavior patterns are viewed as classes defined by regularities in one or more of five domains: location, orientation, physical topography, intrinsic properties, and physical effects. Although the describer selects domains, regularities within domains, and features of regularities, his descriptions are quite objective in a limited but important sense.

Concern over whether behavior patterns are "natural units" is widespread but the term "natural" is variously interpreted and may be too vague to be useful. Various criteria of naturalness are discussed and the conclusion is reached that the complexity of behavior admits of many alternative segmentations. The validity of each should be judged according to the understanding it promotes.

[1] A briefer version of this paper, under the title "Methods of Describing Behavior," was given at the annual meeting of the A.B.S. at Pennsylvania State University in June, 1977.
[2] Present address: Instituto de Biologia, Departamento de Zoología, Universidad Nacional Autonoma de México, Apartado Postal 70-233, México 20 D.F., México.

II. INTRODUCTION

A necessary preliminary to any ethological study is the description of behavior, yet scant attention has been paid to how this proceeds. While considerable emphasis is placed on techniques of quantification and analysis, few authors have concerned themselves with how behavior units are selected and described at the outset. This is particularly surprising since the outcome of any investigation will be profoundly influenced by initial descriptive formulations (Marler and Hamilton, 1966; Hinde, 1970; Golani, 1976).

Behavior patterns are the currency in which ethologists deal, so it is important to have a clear understanding of what they are and how they are described. The present paper attempts to go beyond recent concern with the identity of fixed, or modal, action patterns (Barlow, 1968, 1977; Moltz, 1965; Hinde, 1970; Schleidt, 1974), and provide a framework within which to view all behavior patterns.

The behavior of any organism is a complex system of events and we customarily discover the natural segments, identify those that are instances of the same basic phenomena (the behavior patterns), and then report the common properties of those patterns. We deal with abstractions. Each pattern is a class of events characterized by identity or similarity in one or more of five domains I have labeled physical topography, orientation, location, intrinsic properties, and physical effects.

Behavior is never completely described; the observer may have recourse to a variety of modes of recording and reporting, (e.g., language, photography, sound recording), but all involve some loss or rejection of information (Hinde, 1970, p. 13). Hence it is important to examine the relationship between behavior and description and the operations by which we progress from one to the other, in order to make explicit the role of the describer. One of the conclusions of this analysis is that the description of behavioral phenomena involves the inevitable intrusion of the describer at several levels, that his contribution, however subtle, is open to public inspection, and that the description may nonetheless be quite objective (cf. Simpson, 1973).

III. DESCRIBING BEHAVIOR: TWO METHODS OR ONE?

First we must come to grips with behavior. Psychologists recognize that it is the focus of their discipline, but have been unable to agree on its definition (Handy and Harwood, 1973). This may be attributed to the breadth of phenomena embraced by ordinary usage, and to the diversity of theoretical approaches. However, both ethologists and behaviorists, though frequently divided on explanatory principles, have always recognized implicitly or explicitly that the term encompasses all observable activities of an organism (e.g., Watson, 1924; Skinner, 1931; King and Nichols, 1960). Verplanck (in prep.) includes in his definition the discharge of electric organs, expansion of chromatophores, glandular secretion, and chemical changes producing luminescence. The express inclusion of activities commonly subsumed under physiology is salutary. Frequently we talk of "physiology and behavior" as if they were discrete categories, and conceive of the latter as constituted by movement or muscular contractions. Examples presented below will illustrate how physiological processes are often treated as behavior. An analysis of behavior and its description should encompass the highly diverse gamut of activities and events that comprise our broadest notions of behavior, including sleeping, blushing, electric discharge, immobility, and so on.

Barker (1963) discussed the problem of conceptually separating behavior from the environment in which it occurs. In this paper, which treats only overt behavior and neglects perception, I will argue that behavior is what the animal does and that the integration of organism and environment is such that the descriptions of many patterns must include components of the environment.

Describing should be distinguished from *defining* and *naming*. The first two terms are often used interchangeably, but there is a conceptual distinction. There is also a distinction between a "behavior pattern," a class comprising all those unique behavioral events which are identified as instances of it, and the unique events themselves or "actual behaviors." The distinction is important since we sometimes need to refer to actual behaviors which are not instances of patterns (i.e., behavioral events not belonging to any identified class), particularly in the context of human behavior. Both behavior patterns and actual behaviors can be described but only the former need be defined. In this paper "a behavior" refers to both behavior patterns and actual behaviors, and is also used in a wider sense, as in "reproductive behavior" or "animal behavior."[3] A *description* is a report of the properties of which an actual behavior or pattern is composed. In its most complete ex-

---

I am not following the view that "a behavior" and "behaviors" are ungrammatical synonyms for "behavioral pattern(s)" since in the psychological literature there is abundant precedent for applying these terms to *any* unit of behavior, including patterns, actual behaviors, acts, responses, and such molar units as "seeking attention" and "cooperating." There is value in keeping this general term for broader reference and continuing to apply the term pattern to the restricted class of units ethologists customarily deal with.

pression the reportee is apprised of sufficient information to envisage the behavior perfectly, at least in principle. In the case of an actual behavior this may be achieved by reporting all of its properties (and this can be done exhaustively); in the case of a pattern, the scope of description is limited to those properties that are common to all of its instances, the regularities. The pattern is commonly *defined* by specifying those properties deemed necessary to identify an occurrence or to distinguish it from all other behaviors (cf. Skinner, 1931, 1935). *Naming* is simply applying a term to the behavior, though this itself is fraught with dangers (Marler and Hamilton, 1966, p. 716). A definition may neglect almost all of the information available yet function perfectly well. For example, "head scratching" may be defined as the repeated and rhythmic contact of claws and head. Definitions of this sort, mere definitions, are common in ethological writing, and a great many descriptive studies do not attempt to go further in identifying their patterns. Most descriptions fall somewhere between the extremes of mere definition, which fails to describe, and complete description, replete with redundancy and superfluous detail, their exact location on the continuum being determined by rational and intuitive decisions as well as such factors as the feasibility of nonlinguistic modes such as photography.

Hinde (1970, p. 10, and see Hinde, 1959, 1973) identifies two methods of description: "One involves reference ultimately to the strength, degree, and patterning of muscular contractions (or glandular activity, or change in some other physiological property). The other involves reference not to these changes but to their consequences." He states that these two methods are often used in combination and stresses that they represent different criteria for description, not different types of behavior. His treatment of the disadvantages and advantages of the methods makes it clear that their appropriateness depends on the behavior as well as on the nature of the investigation. This approach, admitting as it does the legitimacy and even necessity of describing some behaviors in terms of environmental consequences, is a substantial improvement on the common view of behavior as (exclusively) muscle contractions, and has been endorsed by other writers (e.g., Hutt and Hutt, 1970; Burghardt, 1973). It is my contention, however, that the two-method dichotomy can be a misleading way of looking at the process of description since it does not acknowledge the prevalence of statements of consequence in almost all descriptions of behavior.

The field observer sometimes uses a narrative style and employs terms from ordinary discourse like "jumping," "climbing a tree," and "sitting on the ground." At this "stream of behavior" level, many statements may refer to either some motor activity *per se* or some environmental consequence, and the behaviors to which they are applied are not described but merely identified or classified as members of some previously recognized category of activity. At another level the behaviors themselves (actual behaviors or patterns) are described or defined, and of course this too is often done in field notes. Hinde's analysis may have some limited application to the first level, which has as much to do with naming as with describing.

The point I wish to make is that patterns of limb and body movement, the primary objects of the first method, involve consequences just as much as behaviors directed at the environment. The new configurations, static or mobile, of an animal's anatomy resulting from a movement are consequences and will almost inevitably be included in a description of the pattern. Movements have topographical consequences. A revealing example is the antipredator response of box turtles, *Terrapene:* The head is drawn back and disappears inside the shell. If that is not description by consequence, then neither is the description of a hermit crab withdrawing into its borrowed shell, nor that of an emerging neonate iguana dropping back into its nest-hole, nor that of a squirrel running up a tree. When movement gives rise to a new relationship between the parts of an animal's body, then that new relationship is just as much a consequence as a new relationship between the animal and its environment. And there is in this respect no difference between describing the defensive behavior of the lizard *Cordylus cataphractus*, which throws its tail forward and grasps it in its jaws so that the whole animal forms a ring (Rose, 1962), and describing a bird scratching its head or a lizard extending its dewlap. In Golani's (1976) application of the Eschkol-Wachmann movement notation to mammalian motor behavior, this distinction between the "trajectory" and consequence of movement is fundamental and is the origin of the insight that consequence may be referred to any of several coordinate systems of which the animal's body is only one.

Thus the description of a behavior pattern may include statements referring to the process of change in position, shape, physiological property, etc. of some part(s) of an organism, or to the consequences of such processes, and statements of the latter kind are included even in descriptions which do not refer to the environment. We may now consider what constitutes a behavior pattern and which kinds of statements are used to describe and define one.

IV. THE DOMAINS OF REGULARITY

If we examine the very broad range of phenomena regarded as behavior patterns, it is apparent that the identity of each one resides in certain regularities, in those properties which are common to all instances, and that the regularities lie within a limited number of domains. I propose therefore that rather than thinking of two methods of description, we think of several domains within which the regularities of patterns lie, and recognize that patterns are necessarily described by stating those regularities. Five domains are recognized in the present analysis: location of the animal in relation to its environment, orientation of the animal to the environment, physical topography of the animal, intrinsic properties of the animal, and physical effects induced in the environment of the animal. In the succeeding paragraphs I hope to show that static states or changes within one or more of these five constitute the essence of all behavior patterns, and consequently that all verbal accounts of patterns, from mere definitions to complete descriptions, must proceed by reporting them. I believe this division into five domains can serve as a useful framework for identifying patterns and understanding their control as well as elucidating the relationship between behavior and description. Naturally, alternative and equally valid analyses are possible. Schleidt and Crawley (1980) have independently developed a framework similar to this one but which collapses the domains into three.

A. Location in Space

In this treatment "location" refers to the fixed or changing *locus* of an animal in relation to some specified component(s) of the environment. Regularity in this domain constitutes a fundamental element of such patterns as "charging," "climbing a tree," "fleeing," and "incubating," and in some descriptions may even stand alone. Nissen (1950) demonstrated that some responses must be described by such reference. In many patterns orientation and location are intimately related, but they can be separated out analytically. For instance, "approaching," defined by change in location, can be achieved with a variety of orientations; an animal may locomote forward, backward, or sideways.

A well-known example of a pattern characterized principally by change in location is the response of the European starling, *Sturnus vulgaris*, to a peregrine falcon. A bird flying in a flock draws closer to its flockmates and the resultant dense mass of birds makes several quick, highly coordinated turns (Tinbergen, 1969).

In describing patterns of this sort great care must sometimes be exercised in specifying the component(s) of the environment in relation to which the spatial change occurs. In some cases, like the queuing behavior of the spiny lobster, *Panulirus argus*, the describer can feel quite safe. In their autumnal mass migrations these animals form queues of up to 65 individuals which maintain continuous locomotion for periods exceeding several hours (Bill and Herrnkind, 1976). Individuals maintain their position behind the next lobster through tactile contact of their antennular rami and first pereiopods with its extended abdomen. Specification of location in this case is a simple matter. The escape response of the herring gull chick, *Larus argentatus*, is more difficult. A few hours after hatching these birds will leave the nest during an alarm and crouch some distance away. Over the next few days the distance increases, and they direct their locomotion toward and into individual shelters (Tinbergen, 1971). Is the change in location away from the nest or toward something? Clearly the alternative descriptions give rise to quite different causal implications.

That modal action patterns (M.A.P.s) may be characterized principally by change in location has been explicitly recognized by Barlow (1968, 1977), who cited the ritualized locomotion paths of fish and bird displays. Barlow used the term "orientation," but within the present framework, the defining property of some of his examples falls clearly within the location domain, consistent orientation (as defined below) being an additional property. It might be less confusing to confine the term M.A.P. (or F.A.P.) to its original use and apply it only to patterns of change in morphology (which are in many cases accompanied by a taxic component), but the behaviors listed above are in any case *behavior patterns*. This is discussed in Section V.

B. Orientation to the Environment

By orientation I mean the disposition of the structures of the animal in relation to some part(s) of the organic and inorganic environment. This includes, but is by no means confined to, the directing of structures toward objects. Descriptive statements within this domain include those detailing the parts of an animal's body which impinge on the environment. An assertion concerning orientation is silent regarding stimulus control though it may give rise to strong presumptions.

In the courtship behavior of the guppy, *Lebistes reticulatus*, there are two displays which are defined substantially by their orientation. For both "leading" and "checking" the male uses the same motor coordination involving the assumption of a sigmoid posture, but in the former the sigmoid is directed away from the female and in the latter it lies perpendicularly across her path (Baerends, 1957).

Barlow (1968, p. 215) indicated that orientation, (comprising, according to this schema, orientation and location), is basic to some M.A.P.s:

> We must be prepared to visualize the C.N.S. counterpart of a F.A.P. as more than an undirected core independent of the environment. The taxic component must be

regarded in many instances as just as centrally determined as the pattern of coordination . . . . What is determined is not necessarily a precise expression of a movement,
but rather a stipulated relation between the organism, its behavior, and key stimuli.

An example of an appetitive behavior pattern in which orientation is of the essence can be found in the foraging behavior of the northern water snake, *Nerodia sipedon* (Drummond, 1979). When "cruising" in search of aquatic prey the snake moves its head across the water surface with the snout and eyes protruding from the surface, to which the head is inclined at an angle of about fifteen degrees. The same orientation of the head is maintained whether the body is resting on the bank, on the water substrate, or even when the snake floats on the water surface. The snake's body is very labile and may assume a very large number of quite different configurations while searching in this way. Its topography is in fact determined primarily by its orientation to the bank, water surface, and substrates. Any instance of the behavior may be described (inadequately) by referring to the topography alone, but description of the *pattern* must rely heavily on reference to orientation since across different environments the latter is constant while the former varies greatly.

Golani (1976) recently demonstrated with the aid of sophisticated computer analysis that many of the movements in the precopulatory interaction of jackals, *Canis aureus*, are determined by the orientation which they bring about; that is, a given limb movement follows the path of the simplest "route of convergence" of the limb segment to a final position defined by either the animal's own body, the partner, or the external environment. Golani termed certain movements which ensure maintenance of an unvarying orientation to a partner "fixations," and when both animals coordinate their movements to maintain a reciprocal fixation it is called a "joint." These movements are defined by orientation and no description that referred exclusively to the animal's body could possibly be adequate.

Here again there is a danger of selecting the wrong components of the environment and making mistaken causal or functional inferences. When the three-spined stickleback, *Gasterosteus aculeatus*, fans its eggs it uses a stereotyped motor coordination and maintains its body at a consistent angle to the horizontal plane of the nest (Baerends, 1957). If the nest is tilted, the fish adjusts its position, to keep constant the angle between the longitudinal axes of its body and the nest. One might conclude that the fish's orientation is controlled by stimuli from the nest, but Baerends displaced the nest horizontally and found that the fish's position (location and orientation) is controlled by several optical beacons, including objects in the environment such as the corner of the tank and to a lesser extent the nest entrance. Fanning *is* characterized by a stereotyped orientation and location relative to the nest, but the stimulus control is another matter.

C. Topography of the Animal

The next domain is the three-dimensional topography of the animal itself. Regularities include those found in some limb movements and static postures, as well as in inflation of body parts, piloerection, and so on, and descriptions of them refer exclusively to individual structures or relate them to other structures. Statements about such regularities comprise the core of the description of many patterns including most M.A.P.s.

Topographical regularity is in fact so basic that it has achieved a dangerous primacy in our assumptions about behavior and description. Some authors have expressed the opinion that the most objective and rigorous descriptions would ideally confine themselves to this domain. That idea probably derives from a narrow concept of behavior as movement *per se* and confusion between describing a behavior pattern as a class and describing the movement component of an actual behavior. An *instance* of a *movement* can be exhaustively described by specifying physical topography in isolation, but many patterns, as classes, can be only very inadequately characterized by the multiplicity of coordinations which comprise their instances.

The tendency to conceive of behavior as no more than movement (the literature abounds with statements to this effect) probably accounts for the neglect of static behavior. Yet the maintenance of a posture is a highly significant activity in the behavioral repertoire of many species. The northern water snake frequently raises its head and anterior trunk above the level of the substrate on which its posterior rests and freezes with its neck bent forward and its head inclined at an angle of about $40°$ above the horizontal. This "peering" posture is normally maintained for several seconds or minutes before the snake rises even higher to assume the posture again, lowers its head to the substrate and starts forward locomotion, or strikes. It is functionally important in visual detection of prey and constitutes an easily recognizable pattern (Drummond, 1980). Altmann (1965) expressly excluded postures from his catalogue of rhesus monkey social behavior because they "fall into a different category" but offered no explanation for this. He did include the adopting of a posture if it seemed to be communicative but this, I suspect, reflects a prejudice against ascribing the status of behavior pattern to a state of immobility. If different activities can precede a posture so that it is not necessarily regarded as part of a larger unit (see Section V), then I can see no reason why it should not be regarded as an independent pattern.

In passing we may note that the conspicuous regularities of most postures lie within the domains of topography and orientation, while those of movements consist of *changes* in topography and/or orientation.

D. Intrinsic Properties of the Animal

Many organisms not only shift their location, control the disposition of their structures, and change shape, but vary certain properties of their integument in discernible ways. Changes in color, temperature, and electrical and chemical properties are observable and properly regarded as behavior. Such changes occur repeatedly in descriptions of patterns and often must do so since they constitute the most important regularities of some. The color patterns of cichlid fishes, for example, change in subtle and complex ways, different patterns serving signal functions (Baerends and Baerends-Van Roon, 1950).

The Indian python, *Python molurus*, broods its eggs by coiling around them and raising its body temperature several degrees (Hutchison et al., 1966). The physiologically mediated raising and sustaining of temperature is just as significant a component of the pattern as the orientation and location. Similarly, the bioluminescent flashing of courting fireflies is the essence of their courtship display patterns. Emission of electric discharge is an integral part of some agonistic display patterns in the Mormyrid fish, *Gnathonemus petersii* (Kramer and Bauer, 1976), and in *Gymnotus carapo* unmodified discharge and discharge arrest function as distinct displays (Black-Cleworth, 1970).

All of the above examples are of behavior patterns involving change in an intrinsic property or the maintenance of a steady state in a property under internal control. There are in addition many patterns whose description requires inclusion of statements concerning colors which do not change. Most birds and mammals cannot effect short-term color changes but nonetheless use color in signaling by exposing colored structures and patches. A great many animals reveal colors serving aposematic and other signal functions. Hence while the color itself is not behavior, it is included in the *description* of behavior patterns to which it is relevant. It is hard to imagine a description of the behavior pattern wherein a butterfly opens its wings to reveal the eyespots that deter its predators which does not mention the eyespots. They are quite as relevant as the firefly's flashes.

E. Changes Effected in the Physical Environment

Many patterns are characterized by the displacement, deformation, division, consumption, destruction, and so on of components of the organic and inorganic environment. All physical changes induced mechanically, chemically, or electrically come within this domain, but not indirect effects such as the response of a conspecific. For instance, displacement of objects is a defining character of all behavior patterns subsumed under "tool-using," and displacement, division, and consumption of food items are the defining characters of many patterns associated with feeding. Description of them can hardly proceed without inclusion of the regularities in physical effects.

This is not to say that the physical effect *is* the behavior. What the animal actually does can be specified by detailing regularities in the other four domains. However, anyone who set out to describe a manipulatory activity like nest-building or digging would be laboring under an enormous handicap if he confined himself to specifying how structures impinge on objects and denied himself the shorthand technique of specifying physical effects. The action of the animal and its effect on the object are two sides of the same coin as long as physical contact between the two is maintained. Physical effects ensuing after termination of contact, such as the passage of excavated earth through the air, are a direct consequence and properly regarded as part of the same behavioral event.

A great many patterns are characterized by another class of physical effects: the extrusion, ejection, deposition, release, etc. of materials. Examples include all patterned behaviors involving pheromone deposition, excretion of waste substances, shedding of integument, and venom injection.

The mechanical disturbance of air or water, whether produced vocally or otherwise, is another physical effect and has been used repeatedly in the microanalysis of behavior patterns (of which it constitutes only a single aspect).

Most patterns involving movement effect physical changes, often of great regularity, which are absolutely inconsequential, at least for the animal in question. Such effects are normally omitted from descriptions. In the same way, regularities lying within the other four domains are frequently ignored. This is discussed at length in Section VI.

F. A Note on Context

Understanding of the function, control, and structure of a pattern is utterly dependent on determination of the relationships between it and its context. By context I mean past, present, and future events in the organic and inorganic environment, including the behavior of the organism under investigation. This is particularly obvious in the case of patterns functioning in intraspecific communication (on this, see Smith 1965, 1969). It does not mean, however, that antecedent and subsequent events are *part of* the pattern. To take a simple example, the threat display of the gorilla (described by Schaller, 1963) includes everything the animal does, including chest beating, throwing, and running sideways, but it does not include the *responses* of the humans or conspecifics to whom the display is directed. Conspecifics (and other animals) are frequently included in the description of patterns, and this is quite clearly necessary when a pattern is defined by change of location or orientation with respect to them.

Thus, while it is acknowledged that patterns must eventually be related to their behavioral and physical context, the analysis presented here is only concerned with the very first stage of ethological enquiry, namely their identification and description. One thing it makes clear is that very often the description of a pattern, even when stripped of contextual information, necessarily includes a component of the environment (in many cases another animal) because the pattern is defined by a spatial relationship between the two.

V. NATURAL UNITS OF BEHAVIOR

A. The Existence of Natural Units

The importance of units to the study of behavior cannot be overestimated. The conventional view is that behavior study must commence with a qualitative survey which picks out the intrinsic segments of the stream of behavior, and it is generally felt that natural units will emerge with extensive observation (Baerends, 1957; Marler and Hamilton, 1966, p. 714; Hutt and Hutt, 1970). However, some authors have drawn attention to an arbitrary element in our unit-making (e.g., Bateson, 1976; Marler and Hamilton, 1966). Are there natural units? And how do we know when we find them? These questions are far from trivial.

It is interesting that behavior ecologists also give considerable importance to the question of units. For Barker (1963, p. 1) these "inherent segments" of the stream of behavior include "alpha waves, psychotic episodes, and games of marbles." Clearly, different types of units are appropriate for different types of study. Indeed, if one considers the enormous complexity of behavioral output and the various domains within which regularities may be discerned, it is apparent that the stream of behavior may be sliced up in numerous ways. The validity of any particular way should be judged according to the increase in understanding it promotes. Those that group unrelated phenomena will quickly prove useless (cf. Skinner, 1935). Patterns of change in physical topography represent only one class of pattern. It was this sort of realization that drove Golani (1976) to describe the movements of golden jackels, *Canis aureus*, and Tasmanian devils, *Sar-*

cophilus harrisii, in three coordinate systems so that statistical procedures could afterward determine whether the regularity of the behavior resided in physical topography or orientation.

Insufficient attention has been given to the different types of behavior unit. Condon and Ogston (1967) found synchrony in patterns of change in speech, body motion, and bioelectric activity as recorded with an E.E.G. machine. Working with behavior which apparently could not be divided into discrete segments, they discovered "process units." Each unit, lasting a fraction of a second, consists of the initiation and sustaining of movement of the body parts in some new direction. The particular body parts and the particular directions are not important: a unit starts when the moving parts of the body change direction and continues until they all cease movement in those directions. These segments of movement synchronize precisely with verbal utterances. Ethologists do not use units of this type, which might indeed be regarded as components of movement rhythms superimposed on and across behavioral patterns. Their validity in the analysis of behavior is vindicated by the discoveries to which their use has led, for example, the "interactional synchrony" between adult speech and the movements of newborn infants (Condon and Sander, 1974), with its implications for linguistic and sociobiological theory. Closer to the other end of the scale are the much more molar units which have been very profitably used by some workers in applied behavioral analysis (e.g., Wahler, 1975; Wahler *et al.*, 1976).

In the light of the above discussion, the question whether there are natural units may be answered quite simply. The behavior of any organism is a complex flux that typically displays considerable harmony or regularity. Once the regularities have been perceived, experience shows that it can be divided up into segments of recurring phenomena, and each one is a natural unit. Ethologists typically seek units whose defining properties fall within the five domains and which are common to a species or some subset thereof. That these patterns exist is an empirical fact. If by "natural" we mean that their existence is not attributable to man's manipulations, then they are no less natural (and no more dependent on human observation) than such units of other biological sciences as leaves, bark, or legs. The fact that the stream of behavior (and trees and animals) can be sliced up in other ways which lead to different insights and understanding in no way detracts from this conclusion. The reader who doubts this conclusion should consult the Blurton Jones (1971) discussion of the advantages of various alternative segmentations of human facial expressions.

Entirely consistent with this approach is the procedure of objectively identifying the natural units by statistical detection and definition of regularity. Dawkins and Dawkins (1973) established that the upstroke of a drinking chick's bill is a natural unit by demonstrating low uncertainty after onset of the movement. It is a unit because its components are so reliably associated. Machlis (1977), tackling a larger unit, attempted with partial success to verify the existence of "bouts" of pecks in chicks by rigorous quantitative analysis of the intervals between acts. Behavior patterns are established in precisely this way, except that regularity is picked out by perceptual processes which carry out a quantifying function (termed "unconscious statistical assessment" by Marler and Hamilton, 1966, p. 715) instead of formal quantification. Leyhausen (1973) pointed out the complexities that must be faced in dealing with variable complex behavior patterns, and proposed three types of "normal behavior": behavioral range, behavioral median, and type behavior.

But some authors have meant something very different by "natural." In 1965, Altmann wrote:

> If one's goal is to draw up an exclusive and exhaustive classification of the animals' repertoire of socially significant behavior patterns, then these units of behavior are not arbitrarily chosen. On the contrary they can be empirically determined. One divides up the continuum of action wherever the animals do . . . . In this sense, then, there are natural units of social behavior. (p. 492, see also Altmann, 1962).

This is quite logical. A unit of *socially significant* behavior is a unit shown to have social significance. Thus any unit, established *prima facie* as a unit by statistical assessment, unconscious or otherwise, is also a natural unit of social behavior when social significance is demonstrated. Of course one must eventually specify operationally what one means by social significance and here there is room for disagreement, but the principle is clear: any unit, established as such by statistical means, is a natural unit within some functional grouping if its functional significance is demonstrated. Making our statistically established units pass through a functional filter confers some legitimacy on them. (For an example of this, see Golani, 1973.)

Another way of legitimizing units is to show unitary causation. It is reasonable to examine proximate causation in the nervous system, but our understanding of the neurological control of behavior patterns is extremely limited. We have very little idea of how neurons interact to coordinate related muscles and no idea of how the nervous system produces different outputs of activity in a muscle in different contexts (Hoyle, 1976). Some progress has been made in tying the firing of individual neurons or networks in the central nervous system with motor output. Research in this area (reviewed by Phillips and Youngren, 1971) has shown that in vertebrates "small motor acts" are more readily elicited by electrical stimulation of the brain than complex patterns, but the situation is far from clear since the same behavior can be elicited by stimulating many different sites, and few sites are tied to only one behavior. Von Holst and Von Saint Paul (1973), in their discussion of E.S.B. and localization of function, pointed out the formidable difficulties to be overcome in identifying sites in the central nervous system that govern complex behavior patterns. Furthermore, environmental stimuli can be very important, suggesting that in some instances E.S.B. merely facilitates these responses. For invertebrates the situation is slightly clearer. By applying a single shock to each of several brain cells of the nudibranch *Tritonia gilberti*, Willows (1967) was able to elicit a complex motor response apparently amounting to a complete M.A.P. The response did not invariably occur and it could also be elicited by termination of a series of shocks, but a firm connection was established between specific cells and a pattern. Does this make the pattern a more legitimate unit? I think we should suspend judgment. If all motor output is generated by the action of the nervous system, then it follows that some correspondence exists between any arbitrarily selected segment of behavior and some event(s) in that system. Is a natural unit of behavior one that is linked to a single nerve cell, one of several identified cells, a group of coacting cells, a ganglion, or some other combination? And what sort of link should it be? The difficulties being encountered at present with the invertebrate command neuron concept (see Davis, 1977) derive from a scarcity of information on the functional relations between nerve cells and the behavior patterns of intact organisms. A synthesis of ethology and neurophysiology is highly desirable. However, in our present state of ignorance concerning the correspondence of their respective subject matters an attempt to force the units of one into the moulds of the other would be misguided. Nonetheless, prospects for research on invertebrates are good, and it should eventually be possible to provide comprehensive documentation of the neurological control of fairly complex behavior patterns (Hoyle, 1976). It may then be possible to derive principles relating levels of neurological functioning to segments of natural behavior. It remains to be seen whether the behavioral units that emerge from such considerations are as useful for the general ethologist as for the neuroethologist. A probable outcome is that some simply defined neuroethological units will be of use to workers in several areas (e.g., neurophysiology, evolution of behavior patterns, ontogeny of behavior patterns) but other nervous-system—behavior correspondences will be sufficiently complex to allow alternative segmentations. That is, the arbitrariness of unit-making will persist. In that situation we would be driven to recognize what is already apparent, namely that neurophysiology is no authority for establishing natural units of *general* validity.

Behaviorists have delimited their units of analysis, response classes, by insisting that every response be controlled by the same class of stimuli (.e.g, Skinner, 1935). Despite the success of this approach in the laboratory, it holds no promise for partitioning the flow of natural behavior because the complexity of field environments and the variability of behavior under multiple stimulus control make precise determination of the correspondence between specific stimuli and segments of behavior exceedingly difficult (see Barlow, 1968).

Credentials of naturalness might also be sought in more remote determinants, but the interaction of gene and gene, and of genotype and environment is such that it is normally futile to look for simple relationships between behavior patterns and their determinants (Manning, 1975; Bateson, 1976). It is far easier to find a causal link between a specific determinant and a feature of a behavior (Manning, 1975; Barlow, 1977; Burghardt, 1977).

In summary, there are quantifiable regularities in the stream of behavior and experience shows we can abstract recurring phenomena and regard them as real units. However, the stream can be divided in many ways and the ethologist's patterns are but one broad class of unit. The validity of any such class may be assessed in terms of the increased understanding to which it gives rise. By choosing to admit only patterns which satisfy specified functional criteria, the investigator can validate their presence in a functional category, but this is done after the candidate patterns have been detected by formal or unconscious statistical assessment. Neurophysiologists may be able to define *their* units of behavior by linking them to neurological events but there is as yet no reason to suppose that the criteria they employ will be valid for all or even most types of ethological study. A search for genetic criteria would be futile, and the tying of patterns to eliciting stimuli presents insurmountable problems.

B. The Role of the Describer

The cases mentioned above of patterns being picked out by formal statistical means are rare. For a long time it has been normal for researchers to present their animal's repertoire of patterns without explaining how it was obtained. Patterns are detected by using what Lorenz (1971) termed the "ratiomorphic" process of Gestalt perception. Unaided observation has its weaknesses, and several writers have pointed out how technical aids and careful analysis may be needed to correct and supplement it (Lorenz, 1971; Eibl-Eibesfeldt, 1972; Schleidt, 1974). It is important to realize that whatever hardware is applied to the task of picking out units the role of the observer as perceiver cannot be eliminated since machines can only make selections they are programed to perform.

No study has been made of the consistency of these perceptual processes across different observers, although some authors have shown concern over the degree to which their units correspond with those of others working independently with the same behavior (e.g., Blurton Jones, 1971; Lipp,

1978, Lipp and Hunsperger, 1978). Barlow (1977) cited primate studies where different observers derived widely different repertoires for the same species, and suggested that greater congruence is usually achieved in studies of lower vertebrates. But is congruence something to be aimed for? Not necessarily. Workers asking different questions may need different units. In the more complex cases it is presumably more valuable to try different segmentations and see which leads to greater understanding.

While the independent arrival of several workers at the same segmentation of behavior may reflect the fact that they share common assumptions and problems, it also suggests strongly that real regularities have been identified: "natural" units have been discovered. However, the real validation of the units awaits the demonstration, which is usually forthcoming, that their use in behavioral analysis promotes understanding. What of the situation where behavior patterns have been defined and several observers achieve good agreement in recognizing them in the stream of behavior? That is, high conventional interobserver reliability is achieved. We may feel convinced that this validates the units, but it does no such thing. It merely shows that real regularities have been picked out — natural units if you will — but it leaves wide open the important question whether they are appropriate to the aims of the study. That is where real validity resides. The danger of misinterpreting interobserver reliability is illustrated in a paper by Blurton Jones (1971), who first asserts that "(interobserver) reliability has been reached on some very large and probably diffuse categorizations of behavior . . . and is clearly no indication of the validity of a behavior item" (p. 365), and after reviewing the congruence achieved by three observers in identifying components of facial expressions claims: "The conclusion from the interobserver tests is that the components are real and valid segments of behavior . . ." (p. 405). Blurton Jones first statement is preferable and in fact is more consistent with his general approach, which is to emphasize agreement of independent segmentations and usefulness in analysis of behavior.

An understanding of the variables affecting categorizations would be especially useful. Behavior ecologists and cognitive psychologists have shown more interest in this and, although they have dealt with a different kind of unit (one that is a mere segment and not a repeated pattern), their results are of interest. Dickman (1963) found that the observer's attribution of purpose or goal affects the way he divides up an episode of human behavior. Newtson et al. (1977) concluded that the stream of human behavior is divided up by a series of breakpoints which themselves are defined by changes in body configuration and ongoing interpretation of the action. His analysis ignores the subject's interaction with his environment (domains 1, 2, and 5) in determining the phenomenal units, but it raises interesting questions for ethologists. Hubbard et al. (1977, cited in Newton et *al.*) found that segmentation differed as a function of prior expectation as to the intent of the characters in the episode. Do ongoing interpretation and theoretical set affect the choice of ethological units? What determines the ethologist's breakpoints? We might also ask whether there is a tendency to notice regularities in some domains at the expense of others. Even if we develop rigorous techniques for the objective determination of stereotypy (e.g., Schleidt and Crawley, 1980) the answers to these questions will still be pertinent, for most ethologists will of necessity go on using their intuition, and all must start with perception.

C. The Level and Scope of the Unit

The principle of hierarchical organization has been with us now for many years and receives reiterated support (Tinbergen, 1969; Baerends, 1957, 1976; Miller *et al.*, 1960; Barker, 1963; Barlow, 1968, 1977; Nelson, 1973; Dawkins and Dawkins, 1973, 1976; Dawkins, 1976; Golani, 1976). Nelson (1973) drew a distinction between hierarchies of embedment, in which behavior units are merely *classified* hierarchically, and hierarchies of connection, in which the hierarchy defines causal relations. Not all authors are willing to espouse the latter as a general principle of behavioral organization, but several types of causal hierarchy have been proposed [e.g., Leyhausen's (1973) "relative hierarchy of moods"], and Dawkins (1976) has argued persuasively that nervous systems probably *should* evolve hierarchical functioning. Nelson suggested that the possibility of "distributed control" be considered, with the corollary that relationships across units be sought through holistic study of behavior. Interestingly in machine intelligence distributed and reversible control are favored over hierarchical systems. If we assume only a hierarchy of embedment, then whatever form the internal dynamics of the system takes, it follows that investigation can proceed at any level or combination of levels and can seek relations within and between levels.

Russell *et al.* (1954) proposed the term "act" for units which cannot be broken down further into components which occur independently. If one act invariably follows another, they are still considered independent unless they only occur in this association. Behavior patterns, then, are deterministic or probabilistic sequences of acts (Slater, 1973), single acts (Barlow, 1977), or clusters of acts occurring simultaneously. But this analysis only takes account of regularities within the domain of physical topography. Series and clusters of acts may enclose or be enclosed within patterns of orientation, location, or intrinsic properties, or run parallel with them. And to complicate matters further, two behavior patterns may be performed at one time.

A criterion is needed for deciding which of the behavioral phenomena prevailing during the execution of a pattern are to be regarded as part of it. Suppose the behavior is, essentially, a stereotyped movement. If that movement is invariably accompanied by a distinctive galvanic skin response and a stereotyped rise in body temperature, are these negligible and functionally insignificant phenomena part of the pattern? If the answer is no, what would be the answer if functional significance were shown, as in the case of physiologically mediated color change accompanying signal movements? Physiological properties which do not change, and movements and postures occurring in parts of the organism distinct from those in which the behavior was perceived should also be considered. A reasonable rule is to include in any pattern all the observable behavioral phenomena which covary with its more salient phenomena. (For an example of a study which examined the association of motor patterns and electric discharge in a fish, see Black-Cleworth, 1970.) Those which do not covary comprise the autobehavioral *context* of the pattern. On this view, the simple tongue-flick pattern of the northern water snake consists of nothing more than specified properties of the moving tongue and occurs in a very wide range of autobehavioral contexts, including lateral undulation and "peering."

Dawkins and Dawkins (1976, p. 739) found that "even stationary parts of the body contribute to a behavior pattern" in the blowfly, *Calliphora erythrocephala*. Transition frequencies between grooming patterns were partially explainable in terms of postural facilitation if the whole animal was considered. There are two equally logical ways of looking at this: (1) the patterns are comprised of the moving limbs only and postural facilitation is an organizational principle ensuring the temporal contiguity of patterns executed in the same autobehavioral context; or (2) each pattern consists of the activities of the whole animal and contiguity tends to prevail between patterns which overlap in their constituents. The second interpretation is consistent with the views presented above.

Eibl-Eibesfeldt and Hass's (1967) discovery that new patterns emerge in the stream of behavior when film is speeded up or slowed down is cause for circumspection. It comes as no surprise that behavior at normal speed is often too fast to follow: we are often painfully aware of our inadequacy in this respect and quickly resort to technical aids. But viewing recordings of behavior at higher speed in order to discern new patterns is less often contemplated (although time-lapse recording is commonly used as a technique to measure such aspects as frequency). Yet the movements of some animals are generally so slow that artificial speeding up is likely to be a profitable recourse. Species will vary in the extent to which the regularities of their behavior fall on a scale that can be effectively scanned by the human observer (cf. the shrew and the sloth), and some activities of any particular species will be more accessible than others (the human eyebrow flash reported by Eibl-Eibesfeldt and Hass went unremarked by scientists for so long presumably because of its rapidity). Note that I am not discussing ability to pick out the details of a pattern but to perceive it in the first place. We can often recognize and identify examples of motor patterns without being able to specify their components: trotting, cantering, and galloping in the horse could not be adequately described until photography made it possible to freeze and scrutinize the movement, but they were identified centuries beforehand.

Detection of patterns is not just a question of perceptual abilities: the sampling period chosen by the field observer, or dictated to him by logistical considerations, will determine which fragments of an animal's behavior go on record and which patterns are revealed.[4] Some well-established patterns have never been observed in their entirety but were discovered by taking widely spaced samples. I am thinking in particular of patterns of long-distance migration in fishes, birds, and turtles. Which brings us to another important sampling factor, the spatial perspective. Just as archaeologists sometimes need an aerial view to pick out the form of remains, so doubtless would similar perspectives often enable ethologists to discern new patterns of animal movement. Such an advantage is mentioned by Eibl-Eibesfeldt and Hass.

Thus regularities in any domain can escape attention if the observer's perceptual bias or sampling technique lead to inadequate coverage of them. It may be that the use of novel sampling techniques (e.g., recording intervals, playback speeds, observation distances) with almost any species will disclose patterns enclosing, enclosed by, or straddling, etc., those that are more readily perceived. Which should be attended to? There is no *a priori* reason why units derived from observations of behavior at natural speeds should be given priority. There is a need here for some exploratory investigation comparing the products of different speeds, etc. Whatever principles emerged from such an inquiry I would argue that units belonging to any speed are "natural" and valid if study of them promotes understanding of the behavior. Indeed, Condon and Ogston's (1967) study already illustrates the point.

---

[4] There are parallels in the study of relations between behavioral events, where choice of sampling interval can affect analysis profoundly (see Hinde, 1970. p. 197, 198; Schleidt, 1973; Beer, 1977).

VI. THE DESCRIPTION OF BEHAVIOR PATTERNS

A. The Selection of Domains

Any description of a pattern must proceed by reporting the regularities and an exhaustive report of all regularities in all domains constitutes a complete description. It is clear from the discussion and examples in Section II that often one domain is more prominent. Thus classic M.A.P.s are characterized by sterotyped changes in physical topography but also frequently incorporate a relatively fixed orientation. The display jump of the damselfish, *Dascyllus marginatus*, on the other hand, is constituted principally of change in location but also of its orientation (and presumably of its motor coordination and effects on the medium), and this too has been classified as a M.A.P. (Barlow, 1977).

Regularity within a particular domain may not be present in the performance of every instance of a pattern. If an animal which characteristically orients its display to a conspecific occasionally performs the same motor coordination when no partner is present, it does not follow that orientation is not a component of the pattern. If it occurs naturally, then the statistical definition of the pattern should express the variability. If it only occurs under some experimental manipulation, for example isolation, the motor coordination denuded of its orientation is an artifact, and should not be viewed as the "true" pattern: it is merely an induced demonstration of a component of it. Patterns which involve manipulation of objects must be similarly treated: deceiving the animal into performing a motor coordination in the absence of its normal object may be very revealing, but it is perverse to think that what it reveals is in some way truer than the normal behavior, given that the pattern can very often be shown to be an adaptation for interacting with that object. This in no way detracts from the usefulness of such experiments, which have contributed numerous insights into stimulus control and neural organization. Some of the classic observations and experiments in the early ethological literature explored the nature and development of the associations between "fixed patterns" of motor coordination and their "taxes" (see Tinbergen, 1969, pp. 81-88).

Whether the term M.A.P. is applied only to motor coordinations, in accordance with the original more restricted concept, or taken to include patterns characterized by orientation and location, like the damselfish display jump, is not central to my argument. The point is that a pattern can be characterized by regularity in other domains, and if Barlow's usage is followed I can see no reason why we should not regard topography, orientation, location, and intrinsic properties as equally viable domains within which to find regularities amounting to M.A.P.s. Furthermore, I believe static behaviors should be regarded as behavior patterns and are candidates for M.A.P. status since they are genuine patterns (see Section IV) and can display considerable constancy of form. Barlow (1977) defined a M.A.P. in terms of movement but intended the term to embrace static displays (p. 103) and would include such behaviors as the prepounce posture of a mousing fox (personal communication). An interesting example is the "peering" posture of the northern water snake, which has a conspicuous form and cannot be regarded merely as the second act in a larger pattern which begins with "rising up" since it can follow orienting as well. Decisions on these questions must await better understanding (and verification) of the M.A.P. category.

Complete descriptions are not normally appropriate; for several reasons it is quite common to omit whole domains. Firstly, some regularities are too obvious to mention: in many instances the orientation of a locomotor pattern is specified only if it is not head first.

Secondly, domains are often considered unimportant. Accounts of locomotor patterns do not normally refer to the disturbance of substrate or medium because these effects are inconsequential. Note, however, that such effects may well be included when they are functional, as when wolves progress over deep snow by bounding in single file and followers profit from the tracks of leaders (Mech, 1970). Similarly, the inclusion of intrinsic properties depends on functional assessment. The color of a fly's legs is not mentioned in a description of its flying, but in describing the escape response of a grasshopper we take care to include the bright flash colors revealed in flight. Physiological activity is treated in the same way: changes in intrinsic properties are ignored when perceived as "means" or neutral by-products rather than "ends." They can, however, be augmented through natural selection and acquire entirely new significance without undergoing qualitative change. Morris (1956), discussing the evolutionary origins of displays, suggested that blushing is derived, through a process of ritualization, from vascular changes originally associated with thermoregulation. Lissman (1958) hypothesized that the electric output of *Gymnarchus* is derived from the change in electric potential that accompanies normal muscle contraction, and described fish which are apparently nonelectric but emit electrical signals showing many of the characteristics of the discharges of electric fishes. Ordinarily no ethologist would dream of including such seemingly peripheral phenomena in a description, but when they achieve functional prominence their inclusion is mandatory. A revealing borderline case is found in the weakly electric fish, *Gnathonemus petersii*. When one of these fishes (which use electric discharge in social signaling) flees from a partner, its characteristic motor behavior is accompanied by a specific discharge pattern (Kramer, 1977). No communicatory function has been established and

Kramer considers the possibility that the signal is an incidental response associated with motor activity or a vegetative reaction. The fact of covariation establishes the electric output as a component of the flight pattern, but its inclusion in a description hangs on the credence given to its possible role in communication.

It is important to appreciate the pervasive role of intuitive assessments of function in determining which domains we report. Such assessments can be so covert and carry such conviction that the describer fails to realize that he is choosing at all. Purton (1978, p. 659) states, in an interesting discussion of function, that it is "plausible to identify, though not define, function on the basis of 'usual effect.'" However, most behavior patterns involve many effects (including effects in the five domains and remoter events like the responses of conspecifics), and whichever definition of function is used, most effects will be excluded in any particular case. From Purton's discussion it is clear that by "usual effect" he really means "those usual effects implicitly judged to be functional." By deemphasizing or overlooking the diverse effects that attend movements of all kinds Purton obscures the observer's role as information selector.

The third class of omissions concerns regularities the observer is unaware of. We do not report regularities we fail to notice, whether because of selective attention or the limitations of our perceptual apparatus. For example, the vibrations produced by the frictional contact of a snake's coils on each other could not normally be perceived, and if detected would undoubtedly be ignored. But those produced by the saw-scaled vipers, *Echis carinatus* and *E. coloratus*, are well within our hearing range (serving, as they do, a defensive function) and are reported (Gans and Maderson, 1973). Electric communication in fishes was, of course, ignored until its discovery in recent years.

B. The Selection of Regularities and Specific Features

Within each domain the describer confines his attention to selected regularities and features. Describing changes in physical topography, he may report only some of the structures involved in the pattern, and he chooses between different measures like velocity, duration, extension in space in one, two, or three dimensions, and frequency. Even reporting a simple intrinsic property like color involves making a choice from hue, brightness, saturation, contrast, duration, etc. The choice is in every case limited to the properties that exhibit regularity and is subject to the same constraints as the selection of domains.

C. The Objectivity of Pure Description

From the foregoing it should be clear that the barest of descriptions is highly subjective. The observer intervenes at every step. A string of qualitative decisions is made as units, domains, and properties are selected, and these decisions inevitably precede objective quantification. Nonetheless, all the data collected can be quite objective since appropriate measurement and statistical treatment will ensure accuracy and verifiability. The point is that objective data based on real regularities can be irrelevant or misleading when judged against the purposes of a study if the wrong selections of information have been made. A description of "cruising" which specified that the snake maintains its eyes just above the water surface and omitted the facts that the nostrils are held out of the water and that the angle and height of the head is such that tongue-flicks strike the water just in front of the snake's jaws, could lead to serious misconceptions but would be quite accurate.

We should, therefore, be aware of the factors influencing our selections. One is the influence that paradigms and hypotheses have on the process of observation. It is easy to imagine how the perception and description of a pattern thought to be a display could be seriously distorted as a consequence of an initial false hypothesis concerning the sensory capabilities of an animal or of its predators or prey, or even of the hypothesis that it is a display. The "perceptual inertia" resulting from initial hypotheses and subsequent "gestalts" has been discussed by Schneirla (1950) and Lorenz (1971), respectively.

Another factor is the parochialism of the human *Umwelt*. And besides psychophysical limitations there are problems of perceiving patterns which exist on a time-scale which places them beyond our unaided powers of resolution (see Section V).

Probably awareness of these factors and of the observer-reporter's role as information-selector is the most important aid to achieving pertinent descriptions. Extensive observation is particularly important in discerning regularities; the common denominators of patterns can be picked out with considerably more assurance if they are viewed repeatedly and in a variety of contexts, and it is often critical that natural settings be included. It is frequently necessary for observation to be intensive, using technical aids which enable reviewing and otherwise overcome our perceptual limitations (e.g., Eibl-Eibesfeldt and Hass, 1967; Eibl-Eibesfeldt, 1972). Refined statistical procedures can reveal patterns beyond our normal computational capacities, and can even serve to distinguish those "feature variables" of a pattern in which variability is lowest (Schleidt and Crawley, 1980). Finally, comparing an animal with a related species can bring our attention to focus on those elements of its behavior whose monotonous recurrence leads us to overlook them, or prepare us to see weaker or incipient versions of a pattern (e.g., Hinde, 1959).

Now, given the most careful observation and reporting, can we ever feel assured that the description of a pattern is uncontaminated with interpretation? In the first place we should remember Schneirla's statement (1950, p. 1036) that "the bare datum of evidence ordinarily regarded as the essence of fact must be recognized as a hypothetical construct." All the entities we construct from raw sensory data, and the data themselves, are constructs in the strict senses intended by Stevens (1935) and Margenau (1950), but (and this is most important) they are in this respect no different from the data of all other sciences, which are also built around the elements of human experience. From this, and the realization that a subjective selection of information necessarily intrudes in the compilation of the most simple behavioral facts, we can conclude, with Schneirla, that *all* descriptions contain interpretive implication. Even so, we can have confidence that our descriptions when taken at face value are quite objective, if we realize that a given pattern may be rendered in many different but valid ways without departing from pure description (cf. McCorquodale and Meehl, 1954). Needless to say, generalization to the population from the sample on which the description is based calls for considerably more circumspection.

Express interpretation is quite another matter. Consider the following statements:

Change X is followed by Y.
The animal does X to produce Y.

The first is purely descriptive, but the second contains an interpretation. Since any change in any domain has consequences (see Section I), it follows that any pattern which is not static can be rendered by statements of either type. The descriptive/interpretive sentences below illustrate both types of rendition:

1. The gill covers are thrust forward and forward locomotion slows immediately.
2. The snake holds its head at $15°$ to horizontal at the water surface with its nostrils in the air.
3. In gaping, young iguanas open their jaws about $30°$ to expose the bright red epithelial lining.
4. By slapping its tail against the water surface with an upward thrust the fish produces a splash.
5. During take-off the downstroke of the wings is extended so that they touch the water surface to form a triangular silhouette when viewed from behind.

The first two are purely descriptive and, by virtue of their selection of information, contain a misleading interpretive implication. The last three include express (and probably erroneous) interpretations whose explicitness varies with the subtleties of the language in which they are couched. In some cases an express interpretation is so obviously correct that none could take exception:

> Prey animals are drawn into the oral cavity by rapid opening of the jaws.

But compare that with the following description of galloping in the horse, which is its formal equivalent:

> Blades of grass and invertebrates are crushed and severed by bringing the hooves sharply down onto the substrate with a rearward thrust.

For an ethologist this interpretation is misleading since it stresses consequences of a pattern which are (presumably) unrelated to its adaptiveness. It is a question of function: the crushing of invertebrates would be included in a description of the pattern if it were suspected that the crushing conferred evolutionarily significant benefits on the individual horse or its kin. On the other hand, the interpretation would be very appropriate for a community ecologist, whose interest in consequences of behavior patterns can be quite different.

Stimulus control presents similar opportunities for deception. In this treatment it is regarded as outside the realm of pure description. Statements detailing consistent orientation, location, or physical effects can be made without any express proposition concerning control. Naturally the statement that a starling moves in the direction of its flock will give rise to very strong inferences regarding stimulus objects and active stimuli, but it is, strictly speaking, noncommittal on those matters. And while stimulus entities may often be safely conjectured, it is as well to bear in mind the complexity of the mechanism governing the "fanning" orientation of the three-spined stickleback (see Section IV).

In conclusion, it is probably unrealistic to attempt to exclude express interpretation from all description since it is often justifiable and saves us from cumbersome language. It is certainly impossible to avoid interpretive implication because it arises from mere selection of information. This is not, however, a cause for despair. Descriptions can be scrutinized and the hypotheses concerning biological significance that they embody can often be tested empirically (see Hinde, 1976, for a discussion of the assessment of function). Awareness of the pitfalls is of paramount importance, for writer and reader, and the former has a heavy responsibility to clarify the boundaries of his own contribution.

VII. CONCLUSIONS

A repertoire of behavior has been found in every animal species in which it has been sought. I do not believe that ethologists need to be at all coy about the status of behavior patterns as legitimate units on which to base a scientific inquiry. Morphological structures are also picked out initially by subjective assessment, and though their inspection is generally facilitated by greater accessibility, the choice of units is similarly complicated by imprecise boundaries and hierarchical organization, and can be just as arbitrary (for a discussion of this, see Greene, 1977).

But we should be very chary about the classification of our units. There is in the literature an implicit concensus that divides patterns into M.A.P.s and "the rest," but the existence of the former as a discrete category is by no means certain. This is at once a sorry and challenging state for a science to be in. The renewed interest in the nature of M.A.P.s and their quantitative investigation is just the sort of development needed, but the scope of the inquiry should be extended to embrace all patterns.

Feelings of insecurity are likely to increase as we move further away from M.A.P.s. These, after all, are the most conspicuous and unambiguous units. It is the apparent amorphousness of the less stereotyped patterns that should attract us, though. Through careful quantitative inquiry it should be possible to uncover their lawfulness and provide descriptions of them that will complement the work being done with M.A.P.s.

In this paper I propose a conceptual structure for the analysis of these less tractable units. I believe it will prove its worth in application to complex behavior patterns of the types performed by the mouthbrooder cichlid *Limnotilapia dardenei* (described in Coeckelberghs, 1976a,b). Motor coordinations are associated with orientation to conspecifics and objects, and regularities of location and physical effect, as well as patterned color changes and states. The degrees of association between the patterns of movement and coloration lend themselves to, or rather demand, an analysis that explores the relationships between domains of regularity and admits they may occur in parallel, overlapping, and enclosing relationships.

The analysis also attempts to come to terms with the describer's function in information selection. A great deal clearly hinges on his assessment of the function of the events under study. The realization that one subjective decision after another determines the selection of information need not lead us to despair of attaining scientific rigor in description. Biological significance is empirically testable, and therefore while we cannot be sure that significant phenomena will not be missed, by exercising proper experimental caution we can ensure that our findings are sound.

VIII. SUMMARY

Ethologists segment the stream of behavior into patterns. Each one is an abstraction, a class comprising all the instances (actual behaviors) that share certain properties. Patterns, but not necessarily actual behaviors, must be defined or described by reporting these common properties (regularities), which are seen to lie in five domains: the *location* of the animal in relation to some component(s) of its environment; its *orientation* (the disposition of its structures) with respect to some component(s); the *physical topography* of the animal; *intrinsic properties* of its integument; and *physical effects* induced in the environment. Every pattern is defined by a regular change or static state in one or more of these domains. Whenever change is involved, description of the regularity may refer to the change *per se* and to its consequences.

This analysis, which extends merely to the simple description and definition of patterns (and does not attempt to embrace description of context) reveals that many patterns are characterized principally by a spatial relationship between animal and environment.

Concern regarding the existence of natural units of behavior has led to two quite different interpretations of the term. According to the first a unit is natural if it has unity of form, that is, if it recurs reliably in that form in ongoing behavior. In practice this is determined by intuitive assessment and (rarely) formal quantification. The second regards a pattern as a natural unit within some functional category if unitary functional significance is shown. However, the candidate units must be identified as such by intuition or quantification before significance is tested, so the second approach in fact assumes the first. Other criteria could be sought, but none is universally satisfactory: (1) an appeal to neurophysiology does not help because there are as yet no rules of correspondence to bridge the gap between the two disciplines, and in any event neurologically defined patterns will be natural units only for neuroethologists; (2) the tying of patterns to controlling stimuli, though it works well with selected responses in the laboratory, is normally impracticable with patterns because of the complexity of stimulus control in natural environments; and (3) a search for genetic criteria would be futile since genes are not linked to most patterns in any simple way, but at best control some aspects of them.

It is argued that the complexity and hierarchical nature of behavior admit of many alternative segmentations. The validity of each should be judged according to the insights and increase in understanding to which it leads. Experience shows that there are units of the sort ethologists call patterns and that investigation based on them has been particularly productive. The process of perception, which is necessarily involved in their detection, remains to be investigated. In identifying a pattern we usually respond to the more salient regularities but these represent only part of the total unit. Every pattern should be taken to include all the behavioral phenomena that covary with its principal properties. The conventional partition of patterns into acts was designed to cope only with physical topography and does not take account of the regularities residing in other domains. Different, but equally valid patterns emerge in the stream of behavior when it is observed with novel sampling techniques, for example with time-lapse and slow motion photography. This phenomenon requires further exploration.

In describing patterns it is customary to omit whole domains of regularity when they are regarded as obvious or functionally unimportant, and when they are simply not noticed. Furthermore, only certain regularities and features from each domain are reported. This selection of information, operating at several levels (units, domains, regularities, and features), can produce (objective) misleading data and provides fertile ground for wrong interpretive inference. Express functional and causal interpretation can be an equally subtle danger, but is at least avoidable. Awareness of the describer's inevitable intrusion in the process of description, and of its precise scope, can mitigate the distortion it might otherwise produce.

The subjective decisions involved in the choice of units parallel those in other sciences. Interpretation of function, probably the most important source of observer bias, is amenable to experimental investigation. Rigorous study of Modal (Fixed) Action Patterns should be complemented with quantitative investigation of patterns showing higher variability. The conceptual framework proposed in this paper is intended as a contribution toward broadening the study of patterns (the basic units of ethology) and explicating the role of the observer in describing them.

IX. ACKNOWLEDGMENTS

This paper is an outgrowth of studies on *Nerodia sipedon* supported by N.S.F. Grant No. BNS-75-02333 to Dr. G. M. Burghardt, and of studies on juvenile iguanas supported by a short-term fellowship from the Smithsonian Tropical Research Institute to the author. During the writing the author was supported by a Hilton Smith Fellowship from the University of Tennessee. I am grateful to Dr. G. M. Burghardt for advice, criticism, and generous provision of facilities, and to Dr. H. W. Greene for valuable feedback and encouragement during discussions. The development of the ideas expressed in this paper owes much to discussions with Dr. W. S. Verplanck and Sylvia Rojas-Drummond. For commenting on the manuscript I wish to thank Drs. G. W. Barlow, J. C. Malone, S. E. Reichert, W. M. Schleidt, and R. G. Wahler. Dr. E. H. Burtt criticized an earlier draft.

X. REFERENCES

Altmann, S.A. (1962). A field study of the sociobiology of rhesus monkeys, *Macaca mulatta. Ann. N.Y. Acad. Sci.* **102**:338–435.

Altmann, S. A. (1965). Sociobiology of rhesus monkeys, II. Stochastics of social communication. *J. Theoret. Biol.* **8**:490–522.

Baerends, G. P. (1957). The ethological analysis of fish behavior. In: Brown, M. E. (ed.), *The Physiology of Fishes*, Vol. II, Academic Press, New York.

Baerends, G. P. (1976). The functional organization of behavior. *Anim. Behav.* **24**:726–738.

Baerends, G. P. and Baerends-Von Roon, J. M. (1950). An introduction to the study of the ethology of cichlid fishes. *Behaviour Suppl.* **1**:1–242.

Barker, R. G. (1963). The stream of behavior as an empirical problem. In: Barker, R. G. (ed.), *The Stream of Behavior*, Appleton-Century-Crofts, New York.

Barlow, G. W. (1968). Ethological units of behavior. In: Ingle, D. (ed.), *Central Nervous Systems and Fish Behavior*, University of Chicago Press, Chicago.

Barlow, G. W. (1977). Modal action patterns. In: Sebeok, T. A. (ed.), *How Animals Communicate*, Indiana University Press, Bloomington.

Bateson, P. P. G. (1976). Specificity and the origins of behavior. In: Rosenblatt, J. S., Hinde, R. A., Shaw, E., Beer, C. (eds.), *Advances in the Study of Behavior*, Academic Press, New York.

Beer, C. G. (1977). What is a display? *Am. Zool.* **17**:153–165.

Bill, R. G., and Herrnkind, W. F. (1976). Drag reduction by formation movement in spiny lobsters. *Science* **193**:1146–1148.

Black-Cleworth, P. (1970). The role of electrical discharges in the nonreproductive social behavior of *Gymnotus carapo (Gymnotidae, Pisces). Anim. Behav. Monogr.* **3**:1–77.

Blurton Jones, N. G. (1971). Criteria for use in describing facial expressions of children. *Hum. Biol.* **43**:365–413.

Burghardt, G. M. (1973). Instinct and innate behavior: toward an ethological psychology. In: Nevin, J. A. and Reynolds, G. S. (eds.), *The Study of Behavior*, Scott Foresman, Glenview.

Burghardt, G. M. (1977). Ontogeny of communication. In: Sebeok, T. A. (ed.), *How Animals Communicate*, Indiana University Press, Bloomington.

Coeckelberghs, V. (1976a). Contribution to the ethology of *Limnotilapia dardenei* (Boulenger, 1899) *(Pisces, Cichlidae).* 1. Description of the elementary actions and the patterns of markings, *Behav. Proc.* **1**:105–124.

Coeckelberghs, V. (1976 b). Contribution to the ethology of *Limnotilapia dardenei* (Boulenger, 1899) *(Pisces, Cichlidae).* 2. Social behavior: qualitative observations. *Behav. Proc.* **1**:125–134.

Condon, W. S., and Ogston, W. D. (1967). A segmentation of behavior. *J. Psychiatr. Res.* **5**:221–235.

Condon, W. S., and Sander, L. W. (1974). Neonate movement is synchronized with adult speech: interactional participation and language acquisition. *Science* **183**:99–101.

Davis, W. J. (1977). The command neuron. In: Hoyle, G. (ed.), *Identified Neurons and Behavior of Arthropods.* Plenum, New York.

Dawkins, R. (1976). Hierarchical organization: a candidate principle for ethology. In: Bateson, P. P. G. and Hinde, R. A. (eds.), *Growing Points in Ethology*, Cambridge University Press, Cambridge.

Dawkins, R. and Dawkins, M. (1973). Decisions and the uncertainty of behavior. *Behavior* **45**:83–103.

Dawkins, R., and Dawkins, M. (1976). Hierarchical organization and postural facilitation: rules for grooming in flies. *Anim. Behav.* **24**:739–755.

Dickman, H. R. (1963). The perception of behavioral units. In: Barker, R. G. (ed.), *The Stream of Behavior*, Appleton-Century-Crofts, New York.

Drummond, H. (1979). Stimulus control of amphibious predation in the northern water snake, *Nerodia s. sipedon. Z. Tierpsychol.* **50**:18–44.

Drummond, H. (1980). Aquatic foraging in some New World natricine snakes: generalists and specialists, and their behavioral evolution. Ph.D. dissertation. University of Tennessee, Knoxville.

Eibl-Eibesfeldt, I. (1972). Similarities and differences between cultures in expressive movements. In: Hinde, R. A. (ed.), *Non-verbal Communication,* Cambridge University Press, Cambridge.

Eibl-Eibesfeldt, I. and Hass, H. (1967). Film studies in human ethology. *Curr. Anthropol.* **8**(5):477–479.

Gans, C. and Maderson, P. F. A. (1973). Sound-producing mechanisms in recent reptiles: review and comment. *Am. Zool.* **13**:1195–1203.

Golani, I. (1973). Non-metric analysis of behavioral interaction sequences in captive jackals, *Canis aureus L. Behavior* **44**:89–112.

Golani, I. (1976). Homeostatic processes in mammalian interactions: a choreography of display. In: Bateson, P. P. G. and Klopfer, P. H. (eds.), *Perspectives in Ethology, Vol. II*, Plenum Press, New York.

Greene H. W. (1977). Phylogeny, convergence, and snake behavior. Ph.D. dissertation. University of Tennessee, Knoxville.

Handy, R., and Harwood, E. C. (1973). A current appraisal of the behavioral sciences. Behavioral Research Council, Great Barrington, Mass.

Hinde, R. A. (1959). Some recent trends in ethology. In: Koch, S. (ed.), *Psychology: A Study of Science, Study 1/2*, McGraw-Hill, New York.

Hinde, R. A. (1970). *Animal Behavior: A Synthesis of Ethology and Comparative Psychology.* McGraw-Hill, London.

Hinde, R. A. (1973). On the design of check-sheets. *Primates* **14**(4):393–406.

Hinde, R. A. (1976). The concept of function. In: Baerends, G., Beer, C. G., Manning, A. (eds.), *Function and Evolution of Behavior*, Clarendon Press, Oxford.

Hoyle, G. (1976). Approaches to understanding the neurophysiological bases of behavior. In: Fentress, J. C. (ed.), *Simpler Networks and Behavior*, Sinauer Associates, Sunderland, Mass.

Hubbard, M., Engquist, G., and Bois, J. (1977). Eliminating perseverance in social judgements: the effect of a second viewing in reversing observers' initial impressions. Unpublished manuscript, University of Virginia.

Hutchison, V. H., Dowling, H. G., and Vinegar, A. (1966). Thermoregulation in a brooding female Indian python, *Python molurus bivittatus. Science* **151**:694–696.

Hutt, S. J., and Hutt, C. (1970). *Direct Observation and Measurement of Behavior*, Charles C. Thomas, Springfield, Ill.

King, J. A., and Nichols, J. W. (1960). Problems of classification. In: Waters, R. H., Rethling-shafter, D. A., and Caldwell, W. E. (eds.), *Principles of Comparative Psychology*, McGraw-Hill, New York.

Kramer, B. (1977). Flight-associated discharge pattern in a weakly electric fish, *Gnathonemus petersii* (*Mormyridae, Teleostei*). *Behavior* **59**:88—95.

Kramer, B., and Bauer, R. (1976). Agonistic behavior and electric signalling in a mormyrid fish, *Ganthonemus petersii. Behav. Ecol. Sociobiol.* **1**:45—61.

Leyhausen, P. (1973). On the function of the relative hierarchy of moods. In: Lorenz, K., and Leyhausen, P. (eds.), *Motivation of Human and Animal Behavior: An Ethological View*, Van Nostrand, New York.

Lipp, H. P. (1978). Aggression and flight behavior of the marmoset monkey *Callithrix jacchus. Brain Behav. Evol.* **15**:241—259.

Lipp, H. P., and Hunsperger, R. W. (1978). Threat, attack and flight elicited by electrical stimulation of the ventromedial hypothalamus of the marmoset monkey *Callithrix jac-chus. Brain Behav. Evol.* **15**:260—293.

Lissman, H. W. (1958). On the function and evolution of electric organs in fish. *J. Exp. Biol.* **35**:156—191.

Lorenz, K. Z. (1971). Gestalt perception as a source of scientific knowledge. In: Lorenz, K. Z. *Studies in Animal and Human Behavior, Vol. II*. Harvard University Press, Cambridge. (First published 1959.)

Machlis, L. (1977). An analysis of the temporal patterning of pecking in chicks. *Behaviour* **63**:1—70.

Manning. A. (1975). Behavior genetics and the study of behavioral evolution. In: Baerends, G. P., Beer, C., Manning, A. (eds.), *Essays in Honor of Professor Niko Tinbergen, F.R.S.*, Clarendon Press, Oxford.

Margenau, H. (1950). *The Nature of Physical Reality*, McGraw-Hill, London.

Marler, P., and Hamilton, W. J. (1966). *Mechanisms of Animal Behavior*, John Wiley & Sons, New York.

McCorquodale, K., and Meehl, P. E. (1954). Edward C. Tolman. In: Estes, W. K. *et al.* (eds.), *Modern Learning Theory*, Appleton-Century-Crofts, New York.

Mech, L. D., (1970). *The Wolf: The Ecology and Behavior of an Endangered Species*, Natural History Press, New York.

Miller, G. A., Galanter, E., and Pribram, K. H. (1960). *Plans and the Structure of Behavior*, Holt, Rinehart & Winston, New York.

Moltz, H. (1965). Contemporary instinct theory and the fixed action pattern. *Psych. Review* **72**:27—47.

Morris, D. (1956). The feather postures of birds and the problem of the origin of social signals. *Behavior* **9**:75—113.

Nelson, K. (1973). Does the holistic study of behavior have a future? In: Bateson, P. P. G., and Klopfer, P. H. (eds.), *Perspectives in Ethology, Vol. I*, Plenum Press, New York.

Newtson, D. Engquist, G., and Bois, J. (1977). The objective basis of behavior units. *J. Pers. Soc. Psychol.* **35**:847—862.

Nissen, H. W. (1950). Description of the learned response in discrimination behavior. *Psych. Rev.* **57**:121—131.

Phillips, R. E., and Youngren, O. M. (1971). Brain stimulation and species-typical behavior: activities evoked by electrical stimulation of the brain of chickens, *Gallus gallus. Anim. Behav.* **19**:757—779.

Purton, A. C. (1978). Ethological categories of behavior and some consequences of their con-flation. *Anim. Behav.* **26**:653—670.

Rose, W. (1962). *The Reptiles and Amphibians of Southern Africa* (2nd edition), Maskey Miller, Cape Town.

Russell, W. M. S., Mead, A. P., and Hayes, J. S. (1954). A basis for the quantitative study of the structure of behavior. *Behavior* **6**:153–205.

Schaller, G. B. (1963). *The Mountain Gorilla: Ecology and Behavior*, University of Chicago Press, Chicago.

Schleidt, W. M. (1973). Tonic communication: continual effects of discrete signs in animal communication systems. *J. Theor. Biol.* **42**:359–386.

Schleidt, W. M. (1974). How fixed is the fixed action pattern? *Z. Tierpsychol.* **36**:184–211.

Schleidt, W. M., and Crawley, J. C. (1980). Patterns in the behaviour of organisms. *J. Social Biol. Struct.* **3**:1–15.

Schneirla, T. C. (1950). The relationship between observation and experimentation in the field study of behavior. *Ann. N.Y. Acad. Sci.* **51**:1022–1044.

Simpson, M. J. A. (1973). Social displays and the recognition of individuals. In: Bateson, P. P. G., and Klopfer, P. H. (eds.). *Perspectives in Ethology, Vol. I.* Plenum Press, New York.

Skinner, B. F. (1931). The concept of the reflex in the description of behavior. *J. Gen. Psychol.* **12**:40–65.

Skinner, B. F. (1935). The generic nature of the concepts of stimulus and response. *J. Gen. Psychol.* **12**:40–65.

Slater, P. J. B. (1973). Describing sequences of behavior. In: Bateson, P. P. G., and Klopfer, P. H. (eds.). *Perspectives in Ethology, Vol. I*, Plenum Press, New York.

Smith, W. J. (1965). Message, meaning, and context in ethology. *Am. Nat.* **99**:405–409.

Smith, W. J. (1969). Messages of vertebrate communication. *Science* **165**:145–150.

Stevens, S. S. (1935). Operational definitions of psychological concepts. *Psychol. Rev.* **42**: 517–527.

Tinbergen, N. (1969). *The Study of Instinct*, Oxford University Press, New York. (Reprint of the 1951 edition, with a new introduction.)

Tinbergen, N. (1971). *The Herring Gull's World*, Harper and Row, New York (First published 1960).

Verplanck, W. S. (in prep.). *A Glossary of Terms*, Irvington, New York. (Revised edition in publication.)

Von Holst, E., and Von Saint Paul, V. (1973). On the functional organization of drives. In: Von Holst, E. (ed.), *The Behavioral Physiology of Animals and Man, Vol. I*, Methuen, London.

Wahler, R. G. (1975). Some structural aspects of deviant child behavior. *J. App. Behav. Anal.* **8**:27–42.

Wahler, R. G., House, A. E., and Stambaugh, E. E. (1976). *Ecological Assessment of Child Problem Behavior: A Clinical Package for Home, School, and Institutional Settings*, Pergamon, New York.

Watson, J. B. (1924). *Behaviorism*, Revised edition, 1959. University of Chicago Press, Chicago.

Willows, A. O. D. (1967). Behavioral acts elicited by stimulation of single identifiable brain cells. *Science* **157**:570–574.

Chapter 2

INDIVIDUAL DIFFERENCES IN ANIMAL BEHAVIOR

P. J. B. Slater

Ethology and Neurophysiology Group
School of Biology
University of Sussex
Brighton BN1 9G6, U.K.

I. ABSTRACT

This article explores reasons for individual differences in animal behavior and points to various ways in which they deserve closer study.
Differences in feeding, mating, or fighting behavior may occur because
selection favors the adoption of different strategies by different individuals.
Variations in signals may arise through selection for animals to be identifiable as individuals or for their relatedness to others to be assessed. The
variability of behavior itself varies between different patterns in which it has
been measured. Variation may arise because the exact form of the behavior
being measured makes little difference from the point of view of selection. It
is also suggested that variability in other cases may come about because, in
an unpredictable environment, the best course of action cannot be forecast.

II. INTRODUCTION

To many people who study animal behavior differences between individuals are a bugbear. Attempts at measurement often reveal high
variance particularly because experience has such a profound molding influence on behavior in subtle ways that are hard to control. The variation
may even be so great as to obscure any effects that experimental treatments
may have unless special efforts are made to reduce it.

The problem of variability is especially acute in field observation, for here it may not be possible to control many aspects of the situation, such as the experience of the individuals being observed and the environmental conditions during observation. At first sight, laboratory studies are easier because the animal can be placed in a more-or-less constant and simple environment in which its courses of action are limited; furthermore, individual variation can be reduced by the use of inbred strains and standard rearing conditions (Slater, 1978). At an extreme, Skinner (1959) recommends such tight experimental control that only trivial differences remain among animals, the main point of studying more than one individual being to establish that such control has been achieved. But the problem with solutions such as these is that they buy consistency at the expense of generalizability (Altmann, 1974). Findings on one strain or in one set of conditions may not apply to others. It is thus a compensation of field-work that, if reliable findings can be obtained despite the relatively uncontrolled conditions, they are likely to have more generality.

No matter what techniques are used to limit variability, results on animal behavior are seldom clear enough to stand out from the data without statistical treatment. The need for statistics arises largely because of the existence of differences among animals in their responses to a particular situation. In many studies, statistical testing is used to extract an average animal from each group and provide justification for ignoring those which behaved differently. It is the main contention of this article that individual differences in behavior should not be treated so lightly and that there are important reasons why they deserve examination in their own right.

Although the work that I shall discuss is primarily ethological, this point has strong implications for other fields of behavior study which involve observations and experiments on animals in more constrained conditions than those traditional in ethology. Whenever the observer places constraints on the animal there is always the possibility that apparently trivial aspects of the intervention are affecting its behavior in unforeseen ways. Lack of variation among animals may give clearcut results, but these may be influenced more by the situation chosen than by the behavioral propensities of the subjects.

Because much ethological work involves observations on animals in relatively unconstrained conditions, it provides a yardstick against which the variability of other results may be judged. To give a basis for such comparisons, and to explore the reasons for variability in behavior, in this article I shall consider some of the situations in which strong individual differences have been found, and theories which would predict their occurrence in others. This helps to indicate a number of respects in which they should be studied more closely, so that the functional significance of variability can be assessed and reasons suggested why some behavior patterns vary among individuals more than others.

III. DIFFERENCES IN FEEDING BEHAVIOR

Feeding behavior is one feature in which individual animals have been found to differ, in some cases in what they eat (Bryan and Larkin, 1972; Harris, 1965; McFarland, 1977), in others in the sorts of places they visit to find it (Partridge, 1976), and in others in the meal patterns that they adopt (Clifton, 1979; Slater, 1974). In a striking instance of the last of these, adult male zebra finches (*Taeniopygia guttata*) were found to have persistent differences in their feeding patterns when isolated and provided with a plentiful supply of seed (Slater, 1974, 1975). Some individuals took meals of a fairly uniform size at more or less regular intervals, the scheduling of their feeding being consistent enough to reveal cycles about 30 min long, one meal taking place in each cycle. The sizes of meals tended to correlate with the lengths of the gaps after them but not with those before, so that when an animal stopped feeding it was possible to predict roughly when the next meal would begin, but it was not possible to predict the size of a meal on the basis of the length of the interval which preceded it. By contrast with these "meal eaters," a smaller number of birds showed a quite different pattern, with feeding bouts which were, on average, shorter, and which showed no regularity in either their timing or length. A further difference was that these birds showed a marked correlation between meal length and that of the preceding gap, but none with the gap after.

Zebra finches can therefore show two contrasting patterns of feeding, a point which would certainly not have come to light if data from different animals had been massed in an effort to obtain a general picture. Splitting the animals into two, rather than more categories was justified on the grounds that there were marked differences between groups but few within them on several different measures. Figure 1 presents a simple model developed to account for the two patterns. Food deficit is seen as rising toward a start threshold when an animal is not feeding and falling to a stop threshold as feeding takes place. There is some inaccuracy in comparing the deficit with these thresholds so that meals do not always start and stop on schedule. Because meals are brief in relation to delays in measurement, it is suggested that the inaccuracy is greatest at the end of a meal (the dotted lines expressing its limits are further apart). If, on this model, an animal starts to feed near the start threshold (Fig. 1A), the pattern generated is similar to that of the meal eaters discussed above, with regular meals and intervals, and meal length correlated with that of the following gap. On the other hand, if an animal starts to feed at random intervals as it comes across food rather than waiting to reach the start threshold (Fig. 1B), the irregular feeding typical of the other group of birds results, and meal size becomes correlated with the length of the preceding interval.

This model provides a plausible account of how minor differences among birds may result in great discrepancies in their feeding patterns. But why do such differences arise in the first place? In a natural population they might be related to the predictability of the food supply, topping up when the animal comes across food being most beneficial if food is erratically dispersed and it cannot be sure when the next feeding opportunity will arise, and regular feeding being most likely where food is widespread and of predictable occurrence. However, this explanation seems more appropriate to differences among populations or species in the wild rather than to those which were found within a laboratory-reared group of the same species. It is

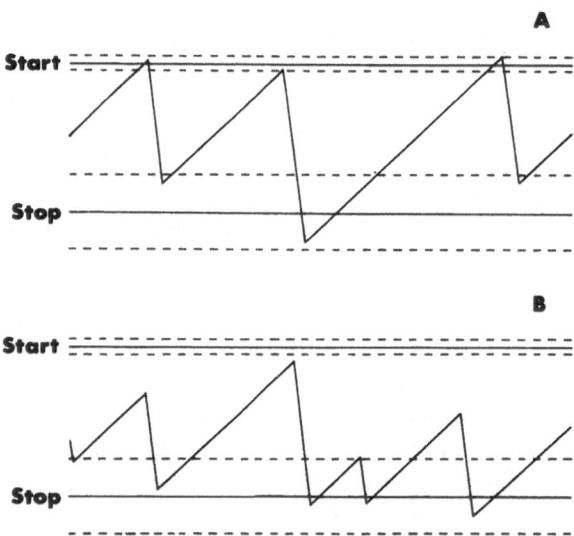

Fig. 1.   A model to account for two different patterns of feeding found in zebra finches. Deficit is seen as rising toward a start threshold during fasting and falling toward a stop threshold during a meal. (A) Birds which eat regular meals are shown as starting and stopping feeding in the region of these thresholds, the limits expressed by the dotted lines. (B) The same model can generate the different pattern shown by irregular eaters if these start to feed at random intervals when they come across food rather than waiting until their deficit rises to the region of the start threshold.

more likely that the animals developed different feeding patterns as a result of the social relationships in the groups in which they were kept, dominant animals having access to the food hopper at will and thus being able to feed regularly (as in Fig. 1A), with subordinates only gaining access when others were not feeding and thus being forced to adopt a less closely scheduled pattern, feeding as the opportunity arose (as in Fig. 1B).

This hypothesis highlights a major reason for the existence of individual differences in behavior: that the behavior adopted by one animal may depend strongly on that shown by others. In many cases behavioral differences among animals may be related to age, sex, or size, but, even where behavior does not correlate with such obvious features, it may be advantageous for some individuals to adopt a different strategy from others. In feeding behavior this has, for example, been shown by Heinrich (1976) in his work on the foraging behavior of bumblebees (*Bombus fervidus*). In this species many individuals feed on common plants with high nectar production so that nectar levels are kept low and each animal has to visit many blooms to get its fill. The few individuals which feed on rarer plants have to fly further between feeds but do just as well because the nectar has more chance to be replenished owing to the lack of competition. Similarly, in seabirds, individuals which feed their young frequently on small fish caught close inshore may have as productive a strategy as those which fly further to catch large fish and can thus only feed their young infrequently (Slater and Slater, 1972).

IV. STRATEGIES OF BEHAVIOR

Interest in the idea that it might benefit individuals to behave differently from each other has increased since Maynard Smith (1972; Maynard Smith and Price, 1973) first used the techniques of game theory to explore problems in animal behavior. These techniques can be applied to any situation in which the behavior best adopted by one animal depends on what others are doing. In some cases it turns out from theoretical analyses and simulations that there is a single strategy (an Evolutionarily Stable Strategy or ESS) which, if adopted by all members of a population, cannot be bettered by any of the other theoretically possible strategies. Thus, a population of "retaliators," animals which display at other individuals when competing over a resource and only fight if their opponent starts to fight, cannot be penetrated by "doves," which only display, or "hawks," which only fight (Maynard Smith and Price, 1973). But in other instances these theoretical arguments predict that two strategies should coexist, neither being able to oust the other from the population, with their proportions determined by the costs and benefits of performing each (Maynard Smith and Parker, 1976).

These models are in many ways unrealistic because they make simplifying assumptions and cannot consider all the possible ways in which animals might behave. They do, however, make predictions about behavior which can be tested, although currently there is a lamentable lack of data on fighting strategies, the area to which game theory has been most fruitfully applied (see the discussion by Caryl, 1979, and Chapter 10 in this volume). The extent to which these predictions should be realized in the form of individual differences remains uncertain. Where the theory suggests that two different strategies should occur in a certain proportion, this balance could be achieved by some individuals adopting one strategy and the rest the other, or by each individual behaving in one or another of the two ways on an appropriate proportion of occasions (Maynard Smith and Parker, 1976). Only with more data on the behavior of individual animals in the wild will it be possible to decide which of these two alternatives does in fact obtain in particular instances.

Sexual strategies have received rather more attention and may, together with foraging behavior, prove a more useful testing ground for models derived from game theory. At present, shortage of detailed quantitative information on the proportions of individuals adopting different strategies, on whether individuals change from one strategy to another, and on the reproductive success that they achieve makes comparison with the models difficult. There is, however, no doubt that there are species in which more than one mating strategy is employed. In the ruff (*Philomachus pugnax*), for example, there are both independent males which set up territories on a lek to which females come for mating, and also satellite males which are tolerated on these territories (Hogan-Warburg, 1966; Rhijn, 1974). Plumage differences between the two types suggest that this behavioral dimorphism has a genetic basis, and individuals do not commonly move from one state to the other. The presence of satellite males on a territory makes it more attractive to females and may therefore increase the mating success of its owner. The satellites do, however, achieve some copulations and are therefore better off than they would be if excluded from the lek.

There are several other cases where males are known to adopt the equivalent of satellite status (Wells, 1977; Le Boeuf, 1974; Perrill *et al.*, 1978). In the green tree frog (*Hyla cinerea*) calling males attract females to their pond, but some males associate themselves with the callers without calling themselves and are often able to intercept approaching females before they reach the male whose call is attracting them (Perrill *et al.*, 1978). Here there must clearly be a balance between the two different types of behavior:

if no individuals called, females would not be attracted, but if all of them did, one or more which adopted peripheral status would achieve more matings than if they devoted time to calling themselves.

V. COMMUNICATION OF IDENTITY

Another reason why individual differences in behavior may arise is that it is beneficial to individuals to communicate some aspect of their identity. Species distinctiveness is common, though not universal (Marler, 1955), among signals. In many cases signals may also incorporate information which makes individual identification possible. Young laughing gulls (*Larus atricilla*), for example, respond only to the "ke-hah" and long calls of their species when these are produced by their own parents (Beer, 1970); in some finches mates adopt similar contact calls, which differ among pairs and enable partners to identify and locate each other (Mundinger, 1970; Marler and Mundinger, 1975).

At a level between species recognition and individual recognition exists the possibility that signals can allow assessment of the degree of kinship between the signaler and a recipient of the signal. This may be true in some bird species in which the young learn components of their song early in life from their fathers (Nottebohm, 1972). Because mistakes are made during cultural transmission, a listener may obtain a rough measure of kinship by comparing a song which it hears with its own or with that of its father: close similarity suggests that a near relative is singing, wide discrepancy that it is a nonrelative. Similar schemes for kin recognition, based on the learning of physical characteristics (Bateson, 1978b) and on smell (Gilder and Slater, 1978), have been suggested in other species.

Interest in the idea that kin recognition may be an important mechanism in social behavior stems from the recent emphasis by evolution theorists on kin selection (Hamilton, 1964), and also on the suggestion that animals should choose their mates so as to achieve an optimal degree of outbreeding (Bateson, 1978a). Kin selection predicts that it may benefit animals to assist their relatives; optimal outbreeding suggests that they should mate with individuals that are not too closely or distantly related to them. The exact ways in which the relationships among animals affect behavior depend on whether individuals provide cues which can be used in assessing kinship. Table I shows three different levels at which kinship recognition may act: which of these is operative in a particular species will have a marked influence on the complexity of the social interactions which it shows. There is, as yet, rather little information enabling one to put different species into one or other of these three categories and, indeed, with some exceptions such as the social insects (Trivers and Hare, 1976) and lions (Bertram, 1976), little is known of the coefficients of relationship among animals in wild populations. Further work is essential to an understanding of the selective forces that have shaped different social structures and of whether or not kinship has had a significant influence.

As Table I shows, animals may be able to behave appropriately toward close relatives by learning to recognize them. But from the point of view of individual differences in behavior the critical question is whether it pays animals to provide cues on which this recognition can be based. In some situations, such as mate choice in a monogamous species, it clearly does as both partners will suffer if they are ill-matched so that offspring of low fitness are produced. In small groups identifiability may have the advantage that it reduces the risk of fighting by indicating to other members that the individual is not an intruder (Bertram, 1970). Individual recognition is also a prerequisite for reciprocal altruism (Trivers, 1972): identifiability may be selected for in species where this is shown because altruistic acts are withheld from unknown animals.

Thus many aspects of the social interactions between animals in groups are likely to be affected by whether or not they can identify each other, either specifically as individuals or less specifically by assessing their degree of relationship. Again, the study of individual differences in behavior emerges as essential if social systems are to be understood more clearly.

Table I. Three Hypothetical Levels of Kin Recognition in Animals and How They May Influence Behavior

1. **No Recognition**
   Animal can only base behavior on probability that other individual is a relative. Kinship can only influence behavior to the extent that this probability varies between species (or sexes) depending on their dispersal patterns.

2. **Individual Recognition**
   Coefficient of relationship (r) may be known exactly, or approximately, for some individuals, but unknown in the case of others. Enables recognition of parent, or sibling with which animal reared, as close relative, but provides no evidence on more distant relations.

3. **Discrepancy Based Recognition**
   Recognition based on variable which shows progressive divergence from value of self or of close kin as r becomes smaller. Possible examples are song in some bird species or smell in mammals. Could give rough measure of kinship over wide range of divergences. May coexist with Category 2.

VI. MODAL ACTION PATTERNS

Signals must vary between individuals if they are to indicate more about the identity of the signaler than simply the species to which it belongs. Individual recognition depends on high variance among individuals but low variance within an individual when it produces the same signal on different occasions (Slater, 1978). It is only in recent years that the variability of both signals and other behavior patterns has been subjected to any scrutiny. This is partly because behavior patterns have often been referred to by ethologists as "fixed action patterns" and regarded as essentially invariant (Schleidt, 1974). The stereotypy of many aspects of behavior, both within and between individuals, is certainly striking. But it is not as marked as has often been made out, and the extent of stereotypy varies a great deal between different actions (Barlow, 1977). Barlow (1968) has suggested the term "modal action pattern" as more appropriate.

The most convenient way of assessing variability is by means of the coefficient of variation (CV), which is the standard deviation of a measure expressed as a percentage of its mean. For morphological features a CV of 2–10% is common (Maynard Smith, 1960), providing a baseline for comparison with behavior. Some behavior patterns show variation lower than this (0–2%), often at the limit of that detectable with the equipment used (Stamps and Barlow, 1973; Wiley, 1973), suggesting that selection has actually favored a high degree of stereotypy. In other cases variation is considerably in excess of this range. Unfortunately, rather few studies have separated the contributions of within- and between-individual variation to the CVs that have been obtained and, here again, there is a need for further work. A wider approach is also called for as most studies involving coefficients of variation have concentrated on signal movements and measures of timing rather than other possibilities (Barlow, 1977).

The potential of this approach to behavioral variability may be illustrated by work on the song of the chaffinch (*Fringilla coelebs*) (Slater and Ince, 1979). Within a population of chaffinches many males may share the same basic song type, having two or three phrases of repeated near-identical notes followed by a complex end-phrase. Some males have only one such song type but others may have up to six (Marler, 1956). Coefficients of variation calculated for several measures are quite different. The intervals between successive notes in two song phrases of a single male, measured in over 50 repetitions, yielded CVs of 1.5% and 3.2%. This variation was probably due largely to measurement error and suggests that selection has favored extreme constancy in this aspect of the song. The number of notes in a phrase shows greater variation. Nine individuals with the same song type

Table II.   Chaffinch Song Repertoire Sizes in Three Different Areas

Number of song types	Individuals with that number in:		
	Cambridge[a]	Orkney[b]	Sussex[c]
1	11	2	2
2	38	5	14
3	16	6	16
4	4	2	8
5	1	1	1
6	1		1
Mean	2.3	2.7	2.9
CV	41.4%	40.1%	35.3%

[a]Marler, 1956.

[b]Slater and Ince, 1979.

[c]Slater and Ince, unpublished data.

yielded CVs in the range 11.6–17.5%. the between-individual CV was 15.6%. These figures show moderate variability, with roughly equal within- and between-individual components, suggesting that this feature has not been selected for stereotypy and also that it is too varied within an individual to be useful for individual recognition. A third aspect is the number of song types which each bird has in its repertoire. Here the results of three studies give CVs which are considerably higher still, ranging from 35.3 to 41.5% (Table II). The reason for this high variability is not yet known, but it certainly does not result from selection for identifiability as the number of song types in a repertoire may take many hours of listening to measure. The existence of such large differences raises a basic question: how much variation in behavior between individuals should one expect without seeking a special explanation in terms of natural selection?

VII. ADAPTIVENESS OR NOISE?

There are two conflicting viewpoints on explaining the variation in the behavior of animals in their natural environment, which may be best illustrated by taking their extreme forms. At one end comes the possibility that a feature, such as repertoire size in chaffinches, may not be relevant from the point of view of selection so that high variance is tolerated. The variation is seen as arising through "noise" in the system, genetic and developmental accidents leading individuals to differ in ways which do not affect their survival and reproductive success. At the other extreme is the view (in keeping with Dr. Pangloss's famous belief that "all is for the best in the best of all possible worlds") that high variation demands explanation and that it would not exist unless it was being actively supported by selection. Where, between these two explanations, the truth lies is a matter for empirical study in each particular case, but the extent to which the adaptation of individual animals to the world in which they live is perfect is certainly a matter of lively controversy (Dawkins, 1976; Lewontin, 1977), the resolution of which is essential to improving our understanding of the functional significance of animal behavior. Although concerned with phenotypic phenomena, this argument is similar to that between neutralist and selectionist explanations of genetic variability (Crow, 1972).

It seems most likely that a moderate degree of variation, such as is common in morphological features, requires no special explanation and can be attributed to noise. As discussed earlier, two possible reasons for greater variability in behavior may be the existence of different strategies or of selection for individuals to be recognizable. Another which deserves consideration is that animals may lack information, from either genetic or environmental sources, about the environment in which they live. Adaptation through genetic change is a slow process which cannot keep up with the rapid fluctuations of an unpredictable environment. Individuals with the capacity to learn are thus at an advantage because, within limits, their behavior can change adaptively during their own lifetimes. But environmental change can still be speedy and unpredictable so that one mode of behavior may be best at one time and others at another, hindering the perfection of adaptation. For example, Lack (1956) found that, in good years for their insect food, swifts (*Apus apus*) which laid three eggs left more young than those which laid two, but in bad years the opposite was the case so that both clutch sizes were maintained in the population. The food supply available later was not predictable at the time of egg-laying so that no mechanism could arise to match clutch size to it. Shortage of information of this sort could also account for variation in behavior in circumstances where the best course of action cannot be forecast with certainty.

As any meteorologist knows, forecasting is an uncertain business even with all the sources of information available to modern man. Animals cannot consult satellite pictures, and their predictions of the weather, as of many other environmental variables, can only be based on the experience of the individual animal through learning and that of its lineage through selection. The behavior of animals can be nicely matched to some aspects of the outside world, such as day length or the form of species specific social signals, because these can be predicted with near certainty. But where the environment changes in a manner which is more probabilistic than deterministic, as must often be the case with such variables as food availability or weather conditions, the behavioral option taken by the individual is more of a gamble, and high variability is to be expected. Selection will clearly encourage such variation, for animals producing offspring which behave consistently in a fluctuating environment, while they may be very successful under some conditions, run the risk of extinction in others.

VIII. CONCLUSION

Individual differences in behavior may thus arise for a number of reasons. Most obviously, it may pay animals to behave differently depending on their age, their sex or their size or on the particular environment in which they find themselves. It may also benefit animals to adopt different strategies depending on what others are doing or to possess signals which vary so as to indicate some aspect of their identity. But variation may also arise either because the exact form of the behavior being measured makes little difference from the point of view of natural selection so that wide limits are tolerated or because the environment varies in an unpredictable manner making it impossible for individuals to do other than gamble on the behavior best adopted. Adaptation can only be perfect, and behavior optimal, where relevant features of the environment are predictable from the information at the animal's disposal. If these features cannot be predicted with certainty, the choice of behavior can only be probabilistic and variations among individuals will arise.

Much of the work that I have discussed concerns ethological studies of the behavior of animals in the wild. Even within this area it is clear that individual differences constitute an important phenomenon in their own right, and I have attempted to point to various areas in which they should be studied further. Individual differences in behavior do, however, have wider relevance. While they are often ignored in laboratory work on behavior, whether psychological or ethological, this has its dangers. Massing animals to obtain a learning curve or a sequence diagram is only useful if there are insignificant differences among them: if not, the average animal which emerges may have a set of features that were not possessed by any single individual in the group. The possibility of animals possessing different strategies or personality profiles can only be explored when they are examined and compared as individuals.

Similar arguments hold at the more functional end of ethology where, as I have tried to show, behavior patterns vary in the extent to which they are subject to individual differences. This variation in variability presents a challenge which is particularly important at a time when interpreting behavior in terms of natural selection is the fashion: does variability exist because it is irrelevant as far as selection is concerned or because selection actively encourages it? Whatever the answer in a particular case, given that variation among individuals is the raw material on which natural selection acts, it should certainly not have been neglected in the way that it has.

IX. ACKNOWLEDGMENTS

I am grateful to R. J. Andrew, P. P. G. Bateson, J. A. Hogan and J. Maynard Smith for helpful comments on earlier drafts of this paper, and to the Science Research Council for financial support.

X. REFERENCES

Altmann, J. (1974). Observational study of behavior: sampling methods. *Behaviour* **49**:227–269.

Barlow, G. W. (1968). Ethological units of behavior. In: D. Ingle (ed.), *Central Nervous System and Fish Behavior*, University of Chicago Press, Chicago, pp. 217–232.

Barlow, G. W. (1977). Modal action patterns. In: T. A. Sebeok (ed.), *How Animals Communicate*, Indiana University Press, Bloomington, pp. 98–136.

Bateson, P. P. G. (1978a). Early experience and sexual preferences, In: Hutchison, J. B. (ed.), *Biological Determinants of Sexual Behaviour*, John Wiley & Sons, London, pp. 29–53.

Bateson, P. P. G. (1978b). Sexual imprinting and optimal outbreeding. *Nature* **273**:659–660.

Beer, C. G. (1970). On the responses of laughing gull chicks (*Larus atricilla*) to the calls of adults. *Anim. Behav.* **18**:652–677.

Bertram, B. C. R. (1970). The vocal behavior of the Indian hill mynah, *Gracula religiosa*. *Anim. Behav. Monogr.* **3**:81–192.

Bertram, B. C. R. (1976). Kin selection in lions and evolution. In: Bateson, P. P. G. and Hinde, R. A. (eds.), *Growing Points in Ethology*, Cambridge University Press, Cambridge, pp. 281–301.

Bryan, J. L., and Larkin, P.A. (1972). Food specialisation by individual trout. *J. Fish. Res. Bd. Can.* **29**:1615–1624.

Caryl, P. (1979). Communication by agonistic displays: what can games theory contribute to ethology? *Behaviour* **68**:136–169.

Clifton, P. (1979). Temporal patterns of feeding in the domestic chick. I. *ad libitum. Anim. Behav.* **27**:811–820.

Crow, J. F. (1972). The dilemma of nearly neutral mutations: how important are they for evolution and human welfare? *J. Hered.* **63**:306–316.

Dawkins, R. (1976). *The Selfish Gene*, Oxford University Press. Oxford.

Gilder, P. M., and Slater, P. J. B. (1978). Interest of mice in conspecific male odours is influenced by degree of kinship. *Nature* **274**:364–365.

Hamilton, W.D. (1964). The genetical evolution of social behaviour. *J. Theoret. Biol.* **7**:1–51.

Harris, M. P. (1965). The food of some *Larus* gulls. *Ibis* **107**:43–53.

Heinrich, B. (1976). The foraging specializations of individual bumblebees. *Ecol. Monogr.* **46**:105–128.

Hogan-Warburg, A. J. (1966). Social behaviour of the ruff, *Philomachus pugnax* L. *Ardea* **54**: 109–229.

Lack, D. (1956). *Swifts in a Tower*, Methuen, London.

LeBoeuf, B. J. (1974). Male-male competition and reproductive success in elephant seals. *Am. Zool.* **14**:163–176.

Lewontin, R. C. (1977). Book review. *Nature* **266**:283–284.

Marler, P. (1955). Characteristics of some animal calls. *Nature* **176**:6–8.

Marler, P. (1956). Behaviour of the chaffinch, *Fringilla coelebs. Behaviour Suppl.* **5**:1–184.

Marler, P., and Mundinger, P. C. (1975).Vocalizations, social organizations and breeding biology of the twite *Acanthus flavirostris. Ibis* **117**:1–17.

McFarland, D. J. (1977). Decision making in animals. *Nature* **269**:15–21.

Maynard Smith, J. (1960). Continuous, quantized and modal variation. *Proc. Roy. Soc. B.* **152**:397–409.

Maynard Smith, J. (1972). *On Evolution*, Edinburgh University Press, Edinburgh.

Maynard Smith, J., and Parker, G. A. (1976). The logic of asymmetric contests. *Anim. Behav.* **24**:159–175.

Maynard Smith, J., and Price, G. R. (1973). The logic of animal conflicts. *Nature* **246**:15–18.

Mundinger, P. C. (1970). Vocal imitation and individual recognition of finch calls. *Science* **168**:480–482.

Nottebohm, F. (1972). The origins of vocal learning. *Am. Nat.* **106**:116–140.

Partridge, L. (1976). Individual differences in feeding efficiencies and feeding preferences of captive great tits. *Anim. Behav.* **24**:230–240.

Perrill, S. A., Gerhardt, H. C., and Daniel, R. (1978). Sexual parasitism in the green tree frog (*Hyla cinerea*). *Science* **200**:1179–1180.

Rhijin, J. G. van (1974). Behavioural dimorphism in male ruffs, *Philomachus pugnax* L. *Behaviour* **47**:153–229.

Schleidt, W. M. (1974). How "fixed" is the fixed action pattern? *Z. Tierpsychol.* **36**:184–211.

Skinner, B. F. (1959). *Cumulative Record*, Appleton-Century-Crofts, New York.

Slater, P. J. B. (1974). The temporal pattern of feeding in the zebra finch. *Anim. Behav.* **22**: 506–515.

Slater, P. J. B. (1975). Temporal patterning and the causation of bird behaviour. In: Wright, P., Caryl, P. G., and Vowles, D. M. (eds.), *Neural and Endocrine Aspects of Behavior in Birds*, Elsevier, Amsterdam, pp. 11–33.

Slater, P. J. B. (1978). Data collection. In: Colgan, P. W. (ed.), *Quantitative Ethology*. Wiley Interscience, New York, pp. 7–24.

Slater, P. J. B., and Ince, S. A. (1979). Cultural evolution in chaffinch song. *Behaviour* **71**:146–166.

Slater, P. J. B., and Slater, E. P. (1972). Behaviour of the tystie during feeding of the young. *Bird Study* **19**:105–113.

Stamps, J. A., and Barlow, G. W. (1973). Variation and stereotypy in the displays of *Anolis aeneus* (Sauria ½ Iguanidae). *Behaviour* **47**:67—94.

Trivers, R. L. (1972). The evolution of reciprocal altruism. *Q. Rev. Biol.* **46**:35—57.

Trivers, R. L., and Hare, H. (1976). Haplodiploidy and the evolution of the social insects. *Science* **191**:249—263.

Wells, K. D. (1977). Territoriality and male mating success in the green frog (*Rana clamitans*). *Ecology* **58**:750—762.

Wiley, R. H. (1973). The strut display of the male sage grouse: a "fixed" action pattern. *Behaviour* **47**:129—152.

Chapter 3

TOWARD A FALSIFIABLE THEORY OF EVOLUTION

Nicholas S. Thompson

Department of Psychology
Clark University
Worcester, Massachusetts 01610

I. ABSTRACT

Over the last two decades, evolutionary theory has been under attack from critics who argue that the "survival of the fittest" is a tautology. The criticism is important because tautological statements cannot be verified empirically. A similar tautology problem existed in the schools of American behaviorism and may have been the cause of their record of meager significant accomplishment. A theoretical program is offered to deal with the tautology in evolutionary theory. The program entails making independent definitions of adaptation and natural selection and then checking to see if natural selection has always begotten adaptation, as the Darwinian theory supposes. A brief review of some well-known biological phenomena reveals that natural selection has often been the cause of phenomena other than adaptation and adaptation has come about through events other than natural selection. The healthy response of a science to such a discovery would be to begin to study the factors other than natural selection which have led to adaptation and to study the consequences of natural selection other than adaptation. Ethology is expected to play a key role in such a response because of its traditional concern with the relationship between comparative descriptive analysis and functional explanation.

II. INTRODUCTION

For several years, evolutionary theory has been under attack from critics who argue that the theory is basically a tautology. The tautology is said to arise from the fact that evolutionary biologists have no widely accepted way to independently define "survival" and "fitness." That the statement, "the fit survive," is tautological is important, because if the critics are correct in their analysis, the tautology renders meaningless much of contemporary evolutionary theorizing. The published argument on the tautology problem has been complex and not well integrated. Authors have not read one another consistently nor understood one another's lines of thought. The problem urgently needs the ministrations of a thoughtful, disciplined, integrative philosopher. But the issue is too central to our field of behavioral evolution for us to wait patiently for the philosophers to work it out.

Consequently, this essay is dedicated to the task of making a start on solving the tautology problem. The solution here proposed is borrowed from the literature on a similar tautology problem which arose among the schools of American behaviorism. The solution makes possible substantial modifications in the contemporary version of the theory of evolution.

III. THE TAUTOLOGY OF EVOLUTIONARY BIOLOGY

To account for patterns of orderliness in biological nature, Darwin made use of a principle of natural selection. Darwin's presentation of his ideas was ambiguous enough to be misconstrued. Since Darwin's time, efforts have been made to clarify the discussion of his theory by rigorously defining its key concepts. Because natural selection was the most precise and mathematically tractable of Darwin's concepts, most evolutionary concepts have come gradually to be redefined in terms of natural selection.

One of the first and most important concepts to be treated in this way was the concept of fitness. In his classical mathematical treatment of natural selection theory, Fisher (1930) redefined fitness as the mathematical complement of natural selection. Fisher also recognized an independent concept, adaptation (1930, pp. 39-41), but he devoted only a few pages to its treatment. But while the distinction may have been clear to Fisher, it has since been disregarded. Thus, for most subsequent authors who have made use of Fisher's conceptualizations, adaptiveness, fitness, and natural selection have become synonymous (e.g., Williams, 1966; Wilson, 1975).

Subsequently, other key concepts have become defined in terms of reproductive success. According to the "biological species concept," two different organisms are thought to be of different species when the selection coefficient against their hybrid offspring is essentially 1.00 (Mayr, 1963, p. 19). The concept "competition" has come to be identified in selection terms because the operational test of whether two organisms are competing is whether the consumption of resources by one produces a detrimental effect upon the reproductive rate of the other. And if competition is defined in terms of natural selection, then so must the ecological niche, since the discovery of overlaps in the niches of two organisms rests ultimately on determining whether they consume the same resources, i.e., are in competition (Mayr, 1963, pp. 67, 78).

The definition of key evolutionary concepts in terms of natural selection runs the risk of making evolutionary theory a self-contained, logical system which is isolated from the empirical world. No meaningful empirical prediction can be made from one side to the other side of these definitions. One cannot usefully predict that nature selects the fittest organism since the fittest organism is by definition that which nature selects. One cannot usefully predict that competition will produce natural selection since competition is by definition that which produces natural selection. One cannot usefully predict that reproductive isolation will produce speciation since speciation is by definition that which reproductive isolation produces. The predictions are useless because they could never be false. The laws of logic — not the laws of nature — require that the fit organism be selected, that selected organisms be competitors and that members of different species be reproductively isolated. Thus, all the tidy relationships among evolutionary concepts are not — as has often been supposed — evidence of the comprehensiveness and solidity of the theory. Rather they are evidence only of determination of evolutionary biologists to define each and every one of their concepts in terms of natural selection. The tragedy is that once the terms are so defined, the theory of evolution can never be said to be true because no method can conceivably be found to prove it false. It has no empirical reference. It is true by definition.

Even if it did not make a tautology of evolution theory, the use of natural selection as a descriptive concept would have serious drawbacks. While it is mathematically tractable and easy to model in the laboratory, the concept is difficult to operationalize in the field. For field biologists, it is really a hypothetical entity. Clear, unambiguous instances of the operation of natural selection are difficult to come by and always greeted with great enthusiasm by biologists (Kettlewell, 1959; Shepherd, 1960). Thus, although the concept has much to recommend it as an *explanatory* one, it seems an overly abstract formulation on which to base a descriptive science.

My understanding, then, of the tautology issue is that the contemporary use of the concept of natural selection runs the risk of making the modern theory of evolution both circular and vague. Circularity means that no empirical information will ever be found that falsifies or limits the central assertion of the theory. Vagueness means that even where predictions are made within the framework provided by the theory, those predictions are going to be difficult to verify or falsify conclusively. The first charge is to the testability of the central theory; the second is directed toward the likely outcome of deciding to take the theory for granted and work within the framework it provides.

The arguments on tautology I have presented here are set forth in a series of debates conducted primarily in contemporary biological journals (MacBeth, 1971; Peters, 1978; Barker, 1969, and references cited therein). Contrary arguments have taken essentially three different forms. Some have argued that modern evolutionary theory serves a classification function and is not exceptional or derelict in being tautological. Others have argued that while being generally tautological, the theory permits falsifiable predictions within limited domains and is productive in this way. A third group has argued that the theory is not tautological if it is properly understood.

The idea that a theory can be both heuristic and nonfalsifiable appears to come from the writings of Thomas Kuhn (1970). According to Kuhn, science is a "puzzle-solving enterprise" which is carried on within the framework of a set of dogmatically held theories. The theories specify the puzzles to be solved and the methods for solving them. An old theory is replaced by a new theory, not necessarily because it is falsified but because the new theory is shown to be a better puzzle-solver. To gain something more substantial than answers to theory-defined puzzles, scientists must make theories which are subject to falsification through experience (Popper, 1968). While Kuhn has presented a very accurate description of scientific activity, his account should not be taken as prescriptive, either for disciplines as a whole or for individual scientists in particular. Scientists should still attempt to produce unambiguous predictions and empirical data to confirm or falsify them. Those who argue that scientists need not strive to make clear and falsifiable predictions from theories are making an inappropriate prescriptive use of Kuhn's descriptive account of scientific progress.

I also take issue with those who have argued that evolutionary theory, although perhaps tautological overall, provides a framework that permits falsifiable predictions. The example often pointed to by those who make this argument (e.g., Ferguson, 1976) is Hamilton's explanation of the broad distribution of eusociality among hymenopteral insects. Hamilton (1964) took note of the fact that hymenopterans have a haplodiploid mating system. In a haplodiploid system, a laying female or queen determines the sex of her offspring: she makes females by fertilizing eggs, males by laying an unfertilized egg. Thus, males have a haploid chromosome number, females a diploid chromosome number. The system makes for unusual genetic relationships among parents and offspring and among sibling offspring. If the queen mates with only one male, sisters of the same mother will have a closer genetic relationship among themselves than they will with their own offspring. Hamilton went on to explain the intense eusociality of many hymenoptera in terms of this odd genetic system, pointing out that many of the details of hymenopteran behavior are consistent with this model.

The example is an excellent one because it demonstrates the perils of working within a tautological framework. The assumption that natural selection is at work does produce testable predictions, such as the prediction that eusociality in insects will always be accompanied by single male mating and haplodiploidy. The prediction is glaringly false. First of all, the eusocial termites are not haplodiploid. Second, the honeybees, on whom Hamilton's theory is based, are eusocial and haplodiploid, but the queen mates with more than one male (Wilson, 1971) with the probable result that females are substantially more closely related to their daughters than to their sisters. Thus, if we take Hamilton's basic notion that social cooperation should parallel the probability of sharing a gene by descent, we would predict that honeybees should be *less* likely to display eusociality than most insects, birds, and mammals. The same conclusion applies to any haplodiploid hymenopterans in which the queen mates with more than two males.

Now obviously there are ingenious lines of thought that can rescue the theory of kin selection from the damaging evidence of honeybee social organization. I do not dispute that the theory *can* be rescued. I even expect that it will. But until the rescuing evidence is developed, the proper course of action is to believe that the honeybees are a negative instance of Hamilton's theory and that *the theory stands provisionally disproven.* But the tautological framework of evolutionary theory has such a grip on the minds of people that negative evidence is actually represented as classical positive evidence for the productivity of the theory. Surely this is doublethink.

The most interesting lines of response to the tautology debate are from those who assert that the theory itself is testable. These writers try to provide an independent assessment of the fitness of an animal, unrelated to the number of its progeny. They provide criteria based on engineering analysis (Gould, 1977, pp. 41-42; Lewontin, 1978, pp. 221-222). Lewontin argues that one can determine the fitness of a zebra for running by comparing its design with that of an ideally modeled zebra. One could then make a prediction that a zebra which is made so as to run faster than another zebra would have better chances of escaping predators and would therefore leave more offspring. In this way, engineering analysis permits us to make non-tautologous predictions from adaptation to natural selection.

As Lewontin recognizes, the problem with using such engineering criteria lies in determining unambiguously the engineering problem to which the organism's design is a response. Such a determination is critical because what may appear to be an imperfect adaptation to one environmental demand may in fact be a perfect adaptation to another. For instance, the great majority of soaring birds have long, straight wings relative to their body length. An engineering analysis will show that this body form gives them the best lift (Pennycuick, 1976). At least one African vulture has fatter, stubbier wings. One might be tempted to leap to the conclusion that this vulture is maladapted. More careful consideration reveals that the fat-winged vulture is doing something slightly different from the thin-winged vultures. While thin-winged vultures are adapted to maximum lift, the fat-winged vultures are adapted for a combination of good lift with high forward groundspeed. They are specialized for beating the other vultures to the carcass (Pennycuick, 1976).

In short, the tautology problem is crucial and unsolved. The argument from design gives some promise of getting us out of the tautology debate, but conceptual progress must be made on the problem of determining the factors to which the organism is adapted. I have my own suggestion for how this might be accomplished. But before I turn to these ideas, I would like to consider a tautology problem that is analogous to evolutionary theory's tautology problem.

IV. THE TAUTOLOGY IN BEHAVIORISM

The science of American behaviorism has had a tautology problem similar to that in evolution theory since its inception almost 70 years ago. Behaviorism's tautology problem surrounds the so-called "Law of Effect" (Thorndike, 1913). The Law of Effect provided that the likelihood of a response by an animal is increased when that response is followed by a satisfying state of affairs. Thorndike himself had a fairly clear idea what he meant by a satisfying state of affairs, i.e., "a set of circumstances which the animal does nothing to avoid and strives to attain" (Thorndike, 1911, p. 245). A satisfier was thus identifiable by studying the behavior of animals to see what circumstances the animal sought. The animal would be expected to learn whatever response immediately preceded the provision of an identified satisfier. For example, if you infer from the behavior of a cat that it is hungry and if you provide the cat with food, whatever the cat was doing just before you provided it with food it will probably do again next time it is hungry. Thus, most of Thorndike's effort was invested in discovering how satisfiers influence behavior, not in discovering how a class of events called satisfiers might be recognized.

A discussion of the tautology issue in behaviorism may well illuminate the discussion of tautology in biology. The structure of the two theories is similar (Postman, 1947). Both assume that the organization of something occurs because directional selection is applied to a set of random responses. In the case of evolutionary theory, the random responses are mutation and recombination; in the case of behaviorism the random responses are the flailings of the untrained animal confined in the strange apparatus. Both theories use a shaping analogy.

As a consequence of this similarity, both theories have sustained attack from opponents who hold that the assumption of randomness in the process that is shaped is gratuitous. Against behaviorism, a school of cognitive behaviorists has held that the animal produces even in a new situation, not random responses, but hypotheses — systematic behavior programs which are more or less adapted to the situation in which the animal finds itself. What the experimenter's reinforcement does is not to mechanistically fit a randomly produced response, but to confirm one of the animal's hypotheses about the nature of the situation he is in (Hilgard, 1948, pp. 274-275). Thus, behaviorism has been under attack from opponents who regard the animal as a much more active participant in his learning process.

Similarly, evolutionary theory has been under attack from a school of neo-Lamarckians. These scientists argue that what is selected by natural selection is not a random mutation, but an organized response of the genome which may be initiated by a random mutation. This line of reasoning is based on the familiar fact that organisms do not react passively to damage during the course of development, but react actively to incorporate the change into ongoing developmental schema of the organism. The results of this assimilation of mutation may or may not be adaptive . . . but it is an organized, not a random response (Piaget, 1971, p. 12).

Given these similarities between the structure and history of the two theories, it is not surprising that behaviorism also has a tautology problem. During the historical period in which Thorndike was thinking up the Law of Effect, it was still respectable to think of animals as having minds. Just as Darwin was not stimulated by the intellectual climate of his time to identify what he meant by fit (or what Spencer meant by fit), Thorndike was not stimulated by the intellectual climate of *his* time to specify what he meant by satisfier. Thorndike was in no doubt as to what constituted a satisfier to the animals he was working with; it would never have occurred to him to wait and see if the cat learned to get out of the box before he concluded that getting out of the box was a satisfier to the cat. His statement, therefore, that satisfying the cat (by letting it out of the box) made more likely whatever the cat was doing just before he satisfied it, was in no sense tautological.

In 1947 a respected student of human learning, Leo Postman (1947), roundly attacked behaviorists for the circularity of their Law of Effect. He argued that unlike Thorndike, most psychologists used the capacity of a situation to produce learning to determine whether or not it was a reinforcer. In reply, Meehl (1950, p. 61) argued that the theory was not circular because the significance of the Law of Effect was not that a certain class of events called satisfiers brings about learning. The real significance of the Law was that a satisfier for the purpose of one learning procedure will prove to be a satisfier for the purpose of other kinds of learning procedures. The Law of Effect was not tautological because it asserted the "trans-situationality" of reinforcers. Since there was no *logical* reason why a reinforcer of one behavior should be a reinforcer of another, the theory was not tautological.

This solution to the problem was widely accepted (Ritchie, 1973, pp. 456–467). It bears some resemblance to the argument in the biological debate that the theory of natural selection permits the making of testable predictions within the benign framework of a tautological theory. The predictions permitted by both arguments are not strictly speaking circular. On the other hand, they fail in both cases to make any statement about the sorts of events that lead to selection on the one hand or reinforcement on the other. They merely assert that those events which have led to reinforcement (selection) in the past are likely to continue to do so. As such, they are convenient instructions for working around a laboratory; but they do not really constitute *a theory* as such.

The issue of tautology in behaviorism lay pretty much dormant until the 1970s, when it was revived by B. F. Ritchie. Ritchie argued that there are at least three routinely observed conditions under which reinforcement does not produce learning. Since these instances are not taken as proving the theory of the Law of Effect false, behaviorists must be guilty of one of two crimes. They are either guilty of tautological thought (the reinforcer ceases to be a reinforcer when it fails to produce learning) or they are guilty of vagueness (the Law of Effect holds only under some set of limiting conditions, but learning theorists have not specified what these conditons are). Ritchie then goes on to suggest an unequivocably nontautological definition of reinforcer and to propose that learning theorists use his definition to settle, once and for all, if and under what conditions reinforcement brings about learning.

If Ritchie is correct in his analysis, then the effect of the tautological Law of Effect upon the conduct and success of American behaviorism has been devastating. During the approximately 60 years since Watson coined the name behaviorism, behaviorists of undeniable skill and intelligence have been at work on the problems of learning. During that period specialists in other fields have split the atom, cracked the DNA code, invented the birth control pill, cloned vertebrate cells, and rocketed to the moon, to name just a few of the momentous scientific accomplishments of our age. During that same period no discoveries of substance have been made by behaviorists. This doleful outcome is summarized by J. A. Deutsch (1960):

Remarkable techniques have been developed for the purpose of collecting data about the behavior of animal organisms under rigorously controlled conditions and for assessing these by elaborate statistical procedure. In this way a large mass of scientifically impeccable evidence has been built up and the pile is daily augmented. We cannot call it an edifice, for in spite of the general agreement about the method of making the bricks, there is no accepted way of putting them together. There is no concord among psychologists about what the facts they have accumulated are evidence for. This does not mean that they are merely in disagreement about the edifice they wish to erect; they have not decided even what constitutes a building. That is, not only do they disagree about the explanation of their findings, but they are not clear about what it could be to explain them. (p. 1)

V. A RESOLUTION OF THE TAUTOLOGY

In the course of his solution to the problem of tautology in the Law of Effect, Ritchie suggests a pattern of thought which might be applicable to the problem of tautology in the theory of natural selection. Ritchie suggests that we state our theoretical assertions in the form of necessity and sufficiency statements. In the case of evolutionary theory this strategy would result in our defining the central theorem of neo-Darwinism as follows:

A history of natural selection is a necessary condition for the adaptation of organisms.

and/or

A history of natural selection is a sufficient condition for the adaptation of organisms.

Having stated the theory in this form, we are sent looking for single instances of a case of adaptation in organisms which was not the result of natural selection and/or single instances of histories of natural selection that have led to conditions other than adaptation. If we find some such instances we declare the theory false as stated.

In such a manner the evaluation of the theory continues until one of two things happens:

1. We find that the need to modify the theory is rarely encountered and we provisionally accept it with slight modification.
2. We find that the modifications of the theory begin to fall into one or more patterns and we decide to supplement or augment the theory with additional principles which incorporate these modifications in an economical way.

My own expectation is that applying this methodology to evolutionary theory will result in significant modifications of the theory. The likely result of our clarifying and detautologizing the theory of evolution will be that the theory will reappear in one or both of the following forms:

A history of natural selection or principle X are necessary conditions to adaptation.

A history of natural selection is a sufficient condition to adaptation or to condition Y.

Now my reader will be saying, all this is very well, but it will fail on just the same point that has defeated all other attempts to write the neo-Darwinian theory of natural selection in nontautological form: the need for an independent criterion for fitness. In reply, I would like to make two assertions which will at first seem outrageous.

Assertion 1. There is in existence a universally accepted definition among biologists of adaptation (fitness) which is entirely independent of natural selection. While the boundary conditions of this definition are vague, it can be (and is) used unambiguously to identify a great many cases of adaptation and of nonadaptation in nature.

Assertion 2. When the definition is rigorously applied it reveals cases in which adaptation has occurred through mechanisms other than natural selection (a history of natural selection is not a necessary condition to adaptation) as well as cases where a history of natural selection has produced consequences other than adaptation (natural selection is not a sufficient condition to adaptation).

The strange thing about these assertions is that while they sound radical, they are simply part and parcel of the ordinary common sense of evolutionary theory. So powerful is the grip of the tautological version of the theory of evolution that we have failed to recognize that it stands provisionally falsified.

Assertion 1

I will first try to justify the assertion that there exists a criterion for adaptation that is universally recognized. It is a strong claim. I base it on the behavior of my colleagues. Evolutionary biologists freely make inferences as to the selective pressures at work in an organism. Since they rarely have any direct evidence of selection at work, they must be making the inference on the basis of something about the organism that is not directly related to natural selection. For instance, all of us might in the course of a casual lecture to an introductory biology course make the assertion that a vulture had been selected by the demands of gliding flight. The point here is not whether that claim is correct; the point is why are we moved to make it? Clearly we do not need to spend dozens of vulture generations counting and classifying baby vultures in terms of the gliding habits of their parents. Something about the *form* of the vulture suggests that it is adapted for gliding flight. Since we take it for granted that natural selection is the cause of adaptation, we work the neo-Darwinian theory of evolution backward and come — quite reasonably — to the conclusion that the vulture was selected by the same demands to which it is clearly adapted. Thus, the evidence of our behavior as scientists suggests that an independent criterion for adaptedness is in use, if only because we make imputations of selectedness so much more often than we have direct evidence of it.

When making inferences of selectedness, scientists appear to use two methods, the comparative method and the modeling method. The comparative method is essentially an elucidation of the concept of natural design. The essence of design is the matching of form to the action which the designed object performs. The designer of buildings, for instance, has in mind a number of forms which a building can take and also a number of activities which go on in buildings. The essence of his designing activity is selecting from among the array of forms he has in mind a building which suits the activities that will go on there. Thus, the appearance of design arose from the matching of items from two arrays, one array being an array of possible forms, the other array being an array of possible activities which the forms may perform. An object is said to be designed when its form, selected from the array of forms, is matched to its actions, selected from an array of possible actions.

Consider entering a new town in an alien culture and trying to decide if the first building we encounter is designed for the activities taking place within it. Let us imagine that the first building we see looks like what we would call a church and contains a group of people doing what we would call worship. Can we say with confidence that this building was designed for worship? No. On the contrary. Our decision will have to wait until we have eliminated three possibilities. First, we will have to eliminate the possibility that all buildings in the town look like churches and that all activities are performed in churches. Then, we have to eliminate the possibility that buildings are of many sorts in this town, but that only one kind of activity goes on in them. We might find people engaged in what we would call worship in buildings which we would call stores, schools, gymnasia, garages, homes, restaurants, etc. Finally, we have to eliminate the possibility that the appropriateness of this particular building to the activities which go on in it is fortuitous and that we will find a variety of activities going on in the town in a variety of buildings but that we will find no consistent matching between building type and activity. Only when we have eliminated these three possibilities can we say with assurance that the first building we have encountered was designed as a place of worship. These possibilities can only be eliminated by taking a fairly large sample of the buildings in our mythical town and of the activities which go on within the buildings and establishing that an improbable degree of covariation exists between building type and activity type. All this information must be gathered before we can state with assurance that the first building we encountered is designed for worship.

This contrived example illustrates that design is a holistic property and that to ask if a particular structure is designed is to seek answers to the following four questions:

1. What is the array of possible forms of the structure?
2. What is the array of activities with which these forms are associated?
3. Is there, in general, a matching of form to activity?
4. Is the particular structure under consideration an example of such matching of form to activity?

The same basic method may be used to recognize design in nature, or as I have defined it, adaptation. For instance, it could be used to determine to what activity and in what degree the beak of an American grosbeak is adapted. One is tempted, because of the fact that grosbeaks use their bills so successfully to crack seeds, to leap to the conclusion that its large heavy beak is adapted to crack seeds. That conclusion is premature; we must first show evidence of *design* for cracking seeds. The array of structures to which the bill of the grosbeak belongs is the array of all bird beaks; the array of activities associated with this structure is the array of all things birds do with their beaks. A study of these arrays reveals a matching between items in the array of beaks and items in the array of activities: stout short beaks tend to be associated with seed-eating and long curved beaks with picking insects from crevices and stout long beaks with hammering against wood, and beaks which are generalized in form with omniverous eating habits. Thus, because of the general correspondence of form with activity in the two arrays, and because the grosbeak is an example of this correspondence, we conclude that the beak is adapted to seed-eating.

The basic outline of this method for assessing adaptation of a particular structure are as follows:

1. Determine the array of variations in the form of the structure.
2. Determine the array of activities associated with the structure.
3. Determine if there exists a correlation between the array of form variation and the array of activities.
4. Determine that the activities associated with the structure under investigation are an example of this correlation.

Once the correlation has been established, we can assess the degree of adaptation by considering how good an example is the structure in question of the correlation we have observed. Those structures which are best adapted are those whose forms correspond most closely to the form typical of structures associated with the activities in question.

The weakness of the comparative method is that it assumes that nature has produced well-adapted organisms with which the subject organism can be compared. An organism is deemed well adapted if its morphology fits its circumstances as well as the organisms whose morphology and behavior are most typical of those circumstances. Yet it is at least logically possible that nature has not yet produced an organism well adapted to those circumstances. One could, for instance, inquire if a particular sort of vulture is well adapted to gliding flight. Assume, for the purpose of argument, that this bird has only the characters of known gliding birds and that conversely all the characters of typical gliding birds are possessed by this species. We would certainly conclude that the bird was well adapted to gliding flight. We would in fact, however, be embarrassed in having made this conclusion if during a subsequent investigation a better glider were discovered. Thus the relative assessment of degree of adaptedness through the comparative method is only as good as nature has been in producing a well-adapted organism. If nature has not been successful in producing a well-adapted organism, then one may attribute a high degree of adaptedness to an animal which subsequent investigation or later process of evolution may reveal to be not so well adapted after all.

A second method, the modeling method, seeks to assess the degree of adaptedness of an organism in a more absolute sense. Once again, the investigator begins by taking account of the morphology and behavior of the organism and of the circumstances under which the organism lives. But now the investigator proceeds in a quite different manner. He decides on what seem to be the critical variables governing the life of an organism under the circumstances of the subject organisms, models those variables with computer or mathematical equations, determines by these means what the optimum physical response would be to the critical variables, and then compares the organism's response to the optimum response in order to assess its relative degree of adaptation. An organism is thought of as being well adapted if its response to the critical variables is close to that calculated as optimal by the investigator's model.

The weakness of the modeling method is that it depends wholly on the quality of the investigator's original intuition concernig the critical variables affecting the organism. An organism is deemed well adapted if it responds to the variables identified by the investigator in an optimal manner, poorly adapted if it responds in a pessimal manner. Yet it is at least logically possible that an apparently poorly adapted organism is in fact perfectly adapted to critical variables other than those identified by the investigator. We could inquire, for instance, if a particular vulture were well adapted by assessing its efficiency as a glider. Assume, for the purpose of argument, that this bird fails to glide as well as the optimal glider identified by our computer program. We would then conclude that the bird is not as well adapted as it might be. However, this conclusion would be a source of embarrassment to us if it were later shown by a subsequent investigation that the bird, while not performing optimally in gliding efficiency, did in fact perform optimally with respect to another variable, say forward ground speed and that forward ground speed was particularly important in this species for securing its food. Thus, the absolute assessment of degree of adaptedness is only as good as the intuition of the investigator in determining in the first place what variables are critical for the organism. If the investigator has guessed at the wrong factors, he may assess a low degree of adaptedness in an animal whose response is optimally fitted to variables other than those that the observer has mistakenly deemed critical.

Because the two methods have complementary weaknesses, they are often combined. The investigator uses the comparative method to determine the most likely critical variables for the subject organism and then uses modeling to determine the subject organism's degree of adaptation to these critical variables. Consider again the hypothetical vulture. Let us say that our comparative investigation reveals that our vulture is not as long-winged as the extreme and that vultures with this less extreme wing shape make their living by a combination of high gliding and high forward ground speed. This analysis suggests that two critical variables for animals of this morphology might be gliding efficiency and forward ground speed. Computer modeling will readily show which wing shape produced the optimum combination of these two factors. Comparison of the subject organism with this ideal will permit an assessment of its degree of adaptation.

In applying this criterion of adaptedness, care must be taken not to change the arrays under discussion during the course of the argument. Otherwise, nonsense is made of the entire enterprise. No matter how aberrant a structure is, it is always adapted to something at some level of analysis. Nobody (I hope) would argue that Franklin Delano Roosevelt was well adapted to the demands of terrestrial life. Certainly he violated the correlation between the demands of a terrestrial existence and the possession of two or more functional legs. Yet in every aspect of his being Franklin Delano Roosevelt displayed evidence of a high degree of adaptation to his leglessness. But such evidence should never be confused with evidence for leglessness being adaptive. The two statements speak of different arrays. The statement that leglessness is well adapted designates an array of locomotory supports (air, water, earth) plotted against an array of species morphologies leading to the conclusion that legs meeting certain design criteria are adapted to use on the terrestrial surface. The statement that F.D.R. was adapted to his leglessness designates an array of legless people of various characteristics plotted against their mobility.

Shifting the arrays under discussion is a great source of mischief in evolutionary thought. For instance, in a discussion concerning the adaptation of species, it is quite inappropriate to speak of the adaptation of members of the species to each other. In the case of discussion of adaptation of species, the relevant arrays are the arrays of species types and the arrays of environments in which the species typically live. For instance, a female red-wing blackbird fits a correlation between ground feeding and nesting birds and the possession of brown-streaked coloration. The male red-wing blackbird does not fit this correlation. The unmistakable conclusion is that the male red-wing blackbird is maladapted for his life style as a terrestrial ground feeding bird. To shift the argument and now say that the red-wing blackbird is adapted to a polygamous mating system is to change the level of analysis. It is like arguing that being legless is an adaptation to political life because Franklin Roosevelt was so marvelously adapted to his leglessness. The fact is that F.D.R.'s legs and the red-wing blackbird's wings violate correlations between habitat and morphology. The fact that these traits are correlated with other characteristics at lower levels of analysis does not make them adapted in the same sense as a good set of legs or a nice streaky brown feather pattern. I will say more on this point later.

Assertion 2

The reader is probably by now beginning to be aware of how it is I have convinced myself that the neo-Darwinian theory of evolution is in need of elaboration and modification. Recall the two statements of the theory with which we agreed to work:

A history of natural selection is a necessary condition for adaptation.

and

A history of natural selection is a sufficient condition for adaptation.

Substituting in our definition of adaptation above, we have:

A history of natural selection is a necessary condition for an organism to be an instance of a correlation between type of organism and type of environment.

and

A history of natural selection is a sufficient condition for an organism to be an instance of a correlation between type of organism and type of environment.

Following Ritchie's pattern of thought, if we find one instance of adaptation which is *not* attributable to natural selection, then the statement that natural selection is a necessary condition for adaptation must be abandoned, modified, or limited in its applicability. Similarly, if we find *one* instance of an outcome of a history of natural selection which is not adaptation, then the statement that natural selection is a sufficient condition for adaptation must also be abandoned, modified, or limited.

Before we can begin to look for such instances we must come to terms with a definition of natural selection. Conceptually, it is not difficult to state a definition of selection to which most biologists will agree: a trait said to be favorably selected when (and only when) bearers of that trait contributed disproportionately to the reproduction of subsequent generations. While conceptually clear enough, this definition is extremely difficult to verify on most species currently known. I suggest the following operational understanding of natural selection:

> A trait may be said to be favorably selected when it can be reasonably attributed to a history of differential reproduction which is consistent with contemporary interactions of the organism with the environment and with other organisms of the same species.

For instance, if we tried to produce definitive evidence that the red epaulettes of the red-wing blackbird were the result of a history of natural selection, we would be in grave difficulties. We can, however, write a reasonable scenario for its selection as a trait and we can show that removal of the epaulettes in a field experiment markedly decreases the effectiveness of a male in maintaining a territory and securing female mates.

Even these relatively loose definitions of adaptation and natural selection make it possible to find several unequivocal exceptions to neo-Darwinian theory. Exceptions to the rule that natural selection always produces adaptation are provided by innumerable instances of traits which are thought to be sexually selected. Consider again the epaulettes of the male red-wing. We have every reason to believe that this characteristic arose as a product of natural selection imposed by the demands of competition among males and/or the demands of females. But the coloration of the male red-wing is *not* adaptive: i.e., it does not obey the correlation between the coloration of the bird and the background against which it lives and feeds.

Most readers will, I think, quickly object that the red epaulettes are an adaptation to the social system of the red-wing or to the predilections of potential mates or competitors. But this line of argument involves the kind of capricious change in arrays proscribed above. In this objection, the red-wings are spoken of as adapting to themselves, whereas the adaptation correlation is a correlation between the form of the animal and the form of its environment. Nothing we know about the environment yields a prediction of red-wings on black background. While the red-wings on a black background are almost certainly a consequence of the mating system and the mating system may be said to be an adaptation to the type of habitat (Orians, 1969), the wing color cannot be said to be an adaptation to the habitat because it obeys no correlation with the environment. The only rigorous conclusion is that natural selection in the case of the male red-wing blackbird has produced a phenomenon other than adaptation.

The same conclusion applies equally to sexual dimorphisms in a great variety of species. Sexual selection does not produce structures which are related by rule to the environment in which the animal lives. Unless we are going to tautologously redefine adaptation as that which natural selection produces, we must now accept the statement that natural selection is not a sufficient condition for the production of adaptation, i.e., that natural selection produces other consequences. We must be concerned with the definition of these other consequences and with some sort of rule for when natural selection produces these consequences and when it produces adaptation. Let us define a "fad":

A trait may be said to be a fad when it is an instance of a correlation between the behavioral preferences of other members of the same species and the form of the trait.

Now our modification of the theory of evolution may proceed as follows:

Natural selection and circumstance X together constitute a condition sufficient for adaptation.

Natural selection and circumstance Y together constitute a condition sufficient for a fad.

The advantage of approaching the problem in this way is that we are led to begin looking for the difference between circumstance X and circumstance Y. Under what circumstances will natural selection produce an adaptation? Under what circumstances will it produce a fad?

Exceptions can also be readily thought of to the proposition that natural selection is *necessary to adaptation*. The common baboon of Africa is widely distributed in a variety of terrestrial habitats ranging from wild savannah to crop land to coastal rock formations (Kummer, 1971). In each location where the baboon is found, he displays food habits appropriate to the region. On coastal rocks, the baboons frolic in the surf and eat shellfish; in cultivated land they raid crops and harass livestock; on savannah land they consume the roots and fruit in season. Unmistakably the food habits of the different populations of baboons are adapted to the local region in which the population lives. Unmistakably, too, this correlation does not come about as a result of natural selection.

The evidence from baboon feeding habits seems to suggest that natural selection is not a necessary condition to the occurrence of adaptation. Most readers will quickly object that the baboons feeding habits are the result of a learning mechanism which is itself a result of natural selection. Once again, the objection entails a shift to a different level of analysis. The adaptation of baboon feeding habits refers to a correlation between the eating habits of baboons and the food supply of their locality. The adaptation of learning mechanisms refers to a correlation between organisms that live in unpredictable environments and the possession of a learning mechanism which determines food habits.

Thus, it makes sense to modify our theory of evolution to fit our knowledge that natural selection is not a necessary condition to the observation of adaptation. Where before we wrote:

Natural selection is a necessary condition to adaptation.

We would now write:

Natural selection *or* a learning mechanism are necessary conditions to adaptation.

We have produced in short order two examples of exceptions to the basic theorems of the contemporary neo-Darwinian synthesis. Are these unique, or could we and should we be on the lookout for others?

I think we should. Once we have a rigorous stable definition of adaptation, we will find that the exceptions are both numerous and interesting. Other places where I suspect we should look for exceptions are as follows:

Eusociality in Insects. Even though the complex social organizations of insects are often cited as examples of the heights of adaptations, they are not adaptations at all in the sense here defined. Eusociality occurs in only one taxonomic group among the insects (Wilson, 1971). Within that taxonomic group, eusociality is widespread and not obviously related to any ecological conditions. This, insofar as current evidence provides, eusociality in insects does not meet the conditions established for the definition of adaptation I have proposed. To the extent that we believe insect sociality to have arisen from natural selection, we must begin to wonder if it is not another exception to the rule that natural selection is a sufficient condition for the observation of adaptation. In a similar manner, I am beginning to distrust the notion that sociality is *ever* an adaptation to the environment. In early attempts a number of authors (Crook, 1968; Jolly, 1972) tried to correlate social organization with ecology in primates with only moderate success. A recent attempt by Harvey and Clutton-Brock (1977) is more successful but seems to involve a shift of the arrays under discussion. The appearance of adaptation is achieved in part by correlating the animal's social organization with other aspects of its behavior, rather than with ecology *per se*.

Brain Size in Mammals. Across a broad number of taxonomic groups there is an increase in relative brain size recorded over evolutionary history (Jerrison, 1973). The trend is so universal that the term adaptation cannot be appropriately applied to it. Similarly warm-bloodedness appears to have a universal utility that crosscuts almost any ecological situation (Bakker, 1971). Trends that are universal cannot properly be called adaptations, although once again we would suspect that natural selection was the cause of these trends. Some innovations may be successful not because they fit the organism more closely to the environment, but because they permit the animal to change the rules of the game entirely. In some sense, such innovations result in the alteration of the environment to fit the needs of the organism. These we might begin to think of not as adaptations to the environment but as assimilations of the environment in the same sense offered by Piaget (1971).

VI. SOME CONCLUDING REMARKS

First, I proposed that evolutionary biology has an unsolved tautology problem which threatens its health as a science. Second, I argued that a parallel tautology has existed in the scientific study of behavior and summarized a line of thought which has been offered by B. F. Ritchie to solve this tautology problem. Finally, I demonstrated how this pattern of thought may be applied to the biological tautology problem with the results that limitations and modifications of the basic theory of adaptation by natural selection are obtained.

Before concluding I would like to step back from these particular tautology problems to ask a broader question: Is there any general lesson to be learned from the fact that we are having tautology problems and what might be gained from learning these lessons?

My impression is that the tautologies develop from attempts to deal with organizational properties of nature. In a first stage of this enterprise, investigators are impressed by the organizational properties of natural events and attempt to find a nonvitalistic explanation for these properties.

In a second stage, a simplifying concept such as natural selection or reinforcement is offered and investigators rush to demonstrate the number of instances of natural organization that this concept accounts for. In a third
stage, investigators become so comfortable with and used to their abstract
explanatory principle that they begin to substitute it for a description of the
thing to be explained. Thus, fitness becomes identified with natural selection, instead of being the thing that is explained by natural selection and
learning identified by reinforcement, not the thing that is explained by reinforcement. This confusion of the thing to be explained with the thing that
explains it probably arises because the abstract explaining constructs —
natural selection and reinforcement — are so much more easily concep-
tualized than the complex organizations of nature that they are designed to
explain. At this point the science is cut off from future growth by its inability
to process information which conflicts with its central theorem.

A solution is to bring our descriptive concepts up to the same level of
precision as our explanatory concepts. A program of description of the
organizational properties of nature is more often proposed than it is carried
out. In his classic work, Williams (1966) urged such a program on biology.
Even though the book is probably the most important single integrative
work in evolutionary biology in the last 20 years, Williams' suggestion has
been mostly ignored. The reasons for Williams' book failing to stimulate
work in the direction he suggested are clear. Williams' work formalizes and
sets in place a tautological definition of adaptation. In short, he defines
adaptation as that which we may reasonably and economically explain by
natural selection. Thus, while Williams himself is acutely aware of the need
for descriptive work, most of his readers become convinced they could use
their concepts of natural selection to identify adaptation in nature. Why
then would anyone go to the arduous and controversial task of describing
organizational properties in nature? No wonder nobody took Williams up!
No wonder he did not take himself up!

In short, the lesson that we must learn from our predilection to
tautology is that we cannot escape the task of describing rigorously the
organizational properties of nature that we are trying to explain. Ethologists
seem to be concerned primarily with two sorts of organizational properties.
The first are those that are revealed by the kind of observational program
which Darwin embarked on when he sailed around the world on the Beagle,
that is, a careful survey of the plants and animals existing in similar environments in different parts of the world. Such an observational program reveals
that species of organisms are matched in their behavior and morphology to
their ecological circumstances. This organizational property is *adaptation*.

The second sort of organizational properties with which ethologists are
concerned are those that are revealed by observing individual animals as

they move through their daily and yearly rounds. This observational program reveals a pattern of relationship between behavior, the circumstances under which it occurs, and its consequences. Such a pattern of relationship corresponds to the concept *motivation*.

If we were able to rigorously define and recognize these two sorts of organizational properties, the science of ethology would be improved. First of all, such definitions would make possible the redefinition in rigorous and straightforward terms of ethology's most troublesome concepts. Such terms as instinct, displacement activities, displays, ritualized activities, and others which are currently defined in terms of ill-defined explanatory entities, could then be recognized as the descriptive concepts that they actually are. Second, the importance of the ethological approach to the evolutionary theory would be made clearer by such a redefinition. The recognition that motivation and adaptation are organizational properties of behavior, not abstract explanatory principles, places a premium on the kind of observation that ethologists do best. Motivation is not revealed except to an observer who knows his animal intimately and over a long period; adaptation is not revealed except to observers who have done extensive comparative studies. The proper study of organizational properties is exactly the kind of study that ethologists are "adapted" to perform.

Third, a definition of these two organizational properties will make possible a consideration of relationships among them. The two properties, adaptation and motivation, exist at different levels of analysis. Yet, much of what ethologists have attempted to do in the last 50 years may be conceived of as an attempt to relate these two organizational properties. The classical ethological paradigm is a demonstration that the goals toward which behavior is directed do not appear to be the same as the functions to which the behavior is adapted. It is this curious disjunction between the motivation and adaptation of organisms that lies at the core of the ethological endeavor.

I conclude that if we make the effort now to exorcise tautological patterns of thought from our science and if we can begin to make rigorous definitions of the complex organizational properties of nature that we hope to explain, that the result will be a reinvigoration of our sense of purpose in the field of ethology.

VII. ACKNOWLEDGMENTS

Pat Bateson has pointed out to me the conceptual debt I owe Niko Tinbergen. The theoretical program outlined in this essay is clearly an abstraction from the empirical program carried out by Tinbergen. Tinbergen regularly makes functional predictions from comparative study of behavior and confirms those predictions with field experiments demonstrating selective factors at work. Unfortunately, not many evolutionary biologists have shared Tinbergen's conceptual clarity nor his profound awareness of the importance of comparative analysis to the study of evolutionary causation. If they had, evolutionary biology would not have had a tautology problem and I would not have needed to write this essay.

I want to thank Rob Peters, Charles Puccia, and Peter Klopfer, good colleagues all, who read earlier versions of this manuscript and helped me improve on it. I also want to thank the members of the Animal Behavior Research Group at Bristol University, in whose conceptual hotbed many of these ideas germinated. Finally, I want to thank Mary Kirby, C. Jaye Shupin, and Becky Clark, whose naive assumption that we could get the manuscript written probably made it possible.

VIII. REFERENCES

Bakker, R. T. (1971). Dinosaur physiology and the origin of mammals. *Evolution* **25**:636–658.

Barker, A. D. (1969). An approach to the theory of natural selection. *Roy. Inst. Phil.* **44**: 271–289.

Crook, J. H. (1968). Ecological and behavioral contrasts between systematic ground-dwelling primates in Ethiopia. *Folia Primatol.* **8**:180–191.

Deutch, J. A. (1960). *The Structural Basis of Behavior*, University of Chicago Press, Chicago.

Ferguson, A. (1976). Can evolutionary theory predict?, *Am. Natur.* **110**:1101–1104.

Fisher, R. A. (1930). *The Genetical Theory of Natural Selection*, Clarendon Press, Oxford.

Gould, S. J. (1977). *Ever Since Darwin*, W. W. Norton and Co., New York, pp. 41–42.

Hamilton, W. D. (1964). The genetical evolution of social behavior I and II, *J. Theoret. Biol.* **7**:1–52.

Harvey, P. H., and T. Clutton-Brock (1977). Primate ecology and social organization, *J. Zool. Lond.* **183**; reprinted in Clutton-Brock and Harvey (1978). *Readings in Sociobiology*, W. H. Freeman and Co., San Francisco.

Hilgard, E. R. (1948). *Theories of Learning*, Appleton-Century-Crofts, New York, pp. 274–275.

Jerison, H. J. (1973). *Evolution of the Brain and Intelligence*, Academic Press, New York.

Jolly, A. (1972). *The Evolution of Primate Behavior*, MacMillan, New York.

Kettlewell, H. B. D. (1959). Darwin's missing evidence, *Sci. Am.* **200**:48–54.

Kuhn, T. S. (1970). *The Structure of Scientific Revolutions*, The University of Chicago Press, Chicago.

Kummer, H. (1971). *Primate Societies: Group Techniques of Ecological Adaptation*, Aldine-Atherton, Chicago.

Lewontin, R. C. (1978). Adaptation, *Sci. Am.* **239**:213–220.

MacBeth, N. (1971). *Darwin Retried*, Dell Publishing Co., New York.

Mayr, E. (1963). *Animal Species and Evolution*, Belknap Press of Harvard University Press, Cambridge, p. 19.

Meehl, P. E. (1950). On the circularity of the Law of Effect. *Psych. Bull.* **47**:61.
Orians, G. H. (1969). On the evolution of mating systems in birds and mammals. *Am. Nat.* **103**:589–603.
Pennycuick, C. (1976). Colloquium presentation before Animal Behavior Round Table, Bristol University, Bristol, England.
Peters, R. H. (1978). Predictable problems with tautology in evolution and ecology, *Am. Nat.* **112**:759–762.
Piaget, J. (1971). *Biology and Knowledge*, University of Chicago Press, Chicago, p. 12.
Popper, Karl R. (1968). *The Logic of Scientific Discovery*, Harper and Row, New York.
Postman, L. (1947). The history and present status of the Law of Effect, *Psych. Bull.* **44**:497.
Ritchie, B. F. (1973). Theories of learning: A consumer report, *Handbook of General Psychology.*
Shepherd, P. M. (1960). *Natural Selection and Heredity*, Harper and Row, New York.
Thorndike, E. L. (1911). *Animal Intelligence: Experimental Studies*, MacMillan, New York, p. 245.
Thorndike, E. L. (1913). *The Original Nature of Man*, Teachers College, New York.
Williams, G. C. (1966). *Adaptation and Natural Selection: A Critique of Some Current Evolutionary Thought*, Princeton University Press, Princeton, N.J.
Wilson, E. O. (1971). *Insect Societies*, Belknap Press of Harvard University Press, Cambridge, Mass.
Wilson, E. O (1975). *Sociobiology*, Belknap Press of Harvard University Press, Cambridge, Mass.

Chapter 4

EVOLUTIONARY, PROXIMATE, AND FUNCTIONAL PRIMATE SOCIAL ECOLOGY

Dennis Robert Rasmussen[1]

Sub-Department of Animal Behavior
University of Cambridge
Madingley, Cambridge CB3 8AA, U.K.

I. ABSTRACT

Social ecology is separated into evolutionary, proximate, and functional subdivisions. In evolutionary social ecology, reconstructions are made of the differences in selection pressures responsible for causing genetically based differences between social organizations. Proximate social ecology is the study of relationships between variation in the immediate environment and variation in social organizations during the lifespan of the members in those social organizations. Functional social ecology is the study of effects of the immediate environment on social organizations and of how those effects are related to the inclusive fitness of individuals within the social organizations. This tripartite division makes it possible to order laboratory, colony, and field investigations of social ecological questions within a single conceptual framework. Consequently, the division may break down some of the barriers between studies of primate social ecology conducted in very different environments. Issues discussed include (1) the necessary link between developmental studies of social behavior and evolutionary social ecology, (2) the current empirical focus on proximate social ecology, (3) the current theoretical focus on evolutionary social ecology, and (4) the need for more studies empirically linking use of environment with social behavior.

[1] Current address: California Primate Research Center, Behavioral Biology Unit, and Department of Psychology, University of California, Davis, California 95616.

II. INTRODUCTION

Social ecology is the study of relationships between variation in environment and variation in social organization (Crook, 1970a, p. 104). In
this chapter, social ecology is separated into evolutionary, proximate, and
functional subdivisions. This tripartite division of social ecology is neces-
sitated by the extremely short lifespan of the scientist relative to the
evolutionary history of primate species. If the social ecologist attempts to
"look" back in time at environmental factors responsible for selecting
genetic bases of differences in social organizations, s/he is concerned with
evolutionary social ecology. If the social ecologist "looks" at effects of the
immediate environment on a social organization during the lifespan of its
members, s/he is concerned with proximate social ecology. If the social
ecologist "looks" at how immediate environmental factors affect the inclusive fitness of individuals by modifying aspects of their social organization, s/he is concerned with functional social ecology.

As within biology as a whole (Tinbergen, 1951, 1963, 1965, 1969), this
tripartite division is for conceptual clarity and focus. In nature all levels are
interdigitated: Selective pressures exerted during the evolutionary history of
a population, or ultimate factors (Baker, 1938, cited by S. Altmann, 1974),
are recorded partially in the genotypes of the currently living individuals
descended from that population. Past selective pressures thus affect
responses of individuals to immediate environmental factors (Pittendreigh,
1958; Struhsaker, 1969). Embedded among immediate environmental factors affecting a population, or proximate environmental factors (Baker,
1938), are ultimate environmental factors that will mold the genetic composition of future populations.

In this chapter, definitions, methods, and issues are discussed in sections on evolutionary, proximate, and functional social ecology. Inter-
relationships among these three subdivisions of social ecology are sketched
in the final section. Attention is focused primarily on primate social ecology
and more on conceptual issues than on reviewing the steadily increasing,
detailed, and carefully executed studies on primate social ecology. Thus, this
chapter provides a perspective on primate social ecology that may aid integration of studies carried out in laboratories, colonies, and the field.

III. EVOLUTIONARY SOCIAL ECOLOGY

Evolutionary social ecology is a historical science concerned with identifying *differences* in selective pressures exerted on ancestral populations

responsible for genetic bases of *differences* between social organizations of extant descendants of those ancestral populations.[2] Evolutionary social ecology is qualitatively distinct from proximate and functional social ecology. Proximate and functional analyses of behavior are potentially empirical sciences, whereas evolutionary reconstructions involving behavior must be speculative (Tinbergen, 1965).

Like all evolutionary studies of behavior (Lorenz, 1950), evolutionary social ecology relies primarily upon the comparative method (e.g., S. Altmann, 1974; Clutton-Brock and Harvey, 1977; Crook, 1970a,b; Crook and Gartlan, 1966; Denham, 1971; Jolly, 1972; Kummer, 1971a). That is, *differences* between individuals or populations are compared with believed *differences* in past selection pressures. While this sounds fairly straightforward, there are major obstacles to be surmounted before strong evolutionary social ecological studies can be conducted.

Whenever selection pressures are invoked as ultimate causes of *differences* in phenotypic traits this is tantamount to asserting that such *differences* have a genetic basis. Thus, in order to support evolutionary explanations of behavioral differences, those differences must be shown to be based on genetic differences:

> The student of evolution must therefore always ensure that the differences he is investigating are in fact indicative of genetic diversity and not merely of dissimilar environments. Special care is needed in behaviour studies because of the plasticity induced by learning processes (Hinde and Tinbergen, 1958, pp. 253–254).

Of course, all aspects of behavior have a genetic basis since a combination of both genetic and environmental factors influences the development of all aspects of behavior, including learning (Garcia *et al.*, 1972; Hinde, 1970; Hinde and Stevenson-Hinde, 1973; Seligman and Hager, 1972; Skinner, 1975; Tinbergen, 1951). Further, all species are genetically different by definition (Mayr, 1963). A superficial concatenation of the last two sentences might lead one to conclude that there will always be genetic differences in the behavioral bases of the social organizations of species. This is not necessarily true. For example, differences between the social organizations of two species might be entirely due to learning (or any other process leading to adaptive modifications[3]) rather than to genotypic differences. Such differences could be said to have a genetic basis, since the

---

[2] I adopt here the traditional and biological use of the terms selection and evolution (Mayr, 1963). Nonrandom genetic alteration of populations occurs as the result of natural selection. Evolution is the consequence of natural selection.

[3] Adaptive modifications are proximally-induced adaptive *differences* among individuals or populations that do not have a genetic basis (Kummer, 1971a,b,c; Lorenz, 1965).

ability to learn must have a genetic basis. Unfortunately, this statement is not the equivalent of asserting that the differences between the social organizations of the two species are due to a genetically based difference in aspects of social behavior of members of those social organizations. This latter assertion and its proof are the very much neglected but essential precursors to strong evolutionary explanations of differences between social organizations.

A computer analogy may help to clarify this point. Suppose two computers are absolutely identical except that one is beige and the other is black. The computers are equipped with identical "hardware" and "software" designed to solve equations that predict how to maximize profits (inclusive fitness)[4] and minimize costs (all factors that decrease inclusive fitness). Because of the different colors of the computers, one is bought by a mercantile firm, and the other by a civil service agency (habitat selection). Over years of use, the computers' output (behavior) and stored information (memories) become very different indeed. Yet the differences between the output and stored information were mediated by their identical "software" and "hardware" and may be traced back to differences in the colors of the computers. It would be quite wrong to state that differences in the computers' predictions of how to maximize profits and minimize costs were in any way due to differences in ways the computers were designed (evolved) to maximize those profits and minimize those costs.

To carry the analogy a step further, let beige computers represent members of a primate species that form groups containing several adult females and several adult males. Black computers represent members of a closely related and recently diverged species that form groups containing one adult male and several adult females. The differences in the social organizations of these two hypothetical species are only due to genetic differences (color) in habitat selection (e.g., Klopfer and Hailman, 1965; Partridge, 1978; Wecker, 1963). It therefore would be a fruitless task to try to reconstruct the differences in the selective agents responsible for genetic differences in social interaction patterns (means of maximizing individuals' inclusive fitness) because there are no such genetic differences. Rather, the evolutionary social ecologist should attempt to reconstruct the selective forces exerted on the ancestral parental population of both species (unpainted computers) responsible for their homologous abilities (hardware

---

[4] Inclusive fitness is the reproductive success of an individual plus the extent to which there is a relative increase in genes it shares in common with individuals who are not offspring (e.g., Hamilton, 1964; Wilson, 1975). Inclusive fitness is discussed further in the section on functional social ecology.

Natural Selection

variation in phenotypes-->some phenotypes have------>increase of the genetic
higher inclusive fitness   material carried by
those phenotypes

Operant Conditioning

variation in behaviour---->some aspects of behaviour--->increase in those aspects
are reinforced   of behaviour

Fig. 1. Similarity of the paradigms of natural selection and operant conditioning. Natural selection results in systematic alteration of gene frequencies in populations. Operant conditioning results in nongenetic changes in behavior. Natural selection operates across individuals and between generations. Operant conditioning works within individuals and within generations. The two paradigms have vastly different time scales. However, in the absence of empirical data, either paradigm may be used with equal facility as a descriptive explanation for differences observed between individuals and their social organizations.

and software) to adaptively modify their social organizations so as to maximize their inclusive fitness.

None of the differences between primate social organizations have been proven to have a genetic basis. A facile evolutionist and a facile learning theorist can therefore readily "explain" all differences observed in social organizations since the paradigms of natural selection and operant conditioning are so similar (Pringle, 1951; Russell, 1959). This similarity is indicated in Fig. 1. A major handicap faced by early learning theorists was the circularity of the definition of reinforcement. This circularity may be readily avoided. For example, a modern theorist could assert that those things which are found to be positively reinforcing to subjects are those which were positively related to the inclusive fitness of ancestors of those subjects.[5] With such a formulation, the learning theorist may acocunt for differences in primate social organizations and not invoke selection for, and genetic differences in, tendencies toward one form of social organization or another. On the contrary, genetic similarity in learning abilities is postulated to be responsible for *differences* in social organizations.

The similarities of the paradigms of natural selection and operant conditioning have been pointed out to illustrate the rather large differences in

[5] Skinner (1975, 1976) seems to be approaching such a formulation, and this formulation is implicit in Boehm (1978) and Trivers (1974).

conclusions that could be drawn by social ecologists with strong attachments to one paradigm or the other. The value of the paradigms is not in question. Both natural selection and operant conditioning have been experimentally demonstrated to be valid. What is in question is the extent to which natural selection, operant conditioning, and other processes actually have been responsible for observed differences in social organizations; a crucial issue in any evolutionary reconstruction.

All explanations of differences in primate social organizations to date have been based upon the descriptive use of the paradigms natural selection and operant conditioning. Thus either evolutionary explanations or learning explanations of species' differences in social organizations are built upon the shifting sands of scientific belief and not yet upon empirical bedrock. At one point Crook (1970b, p. 198) urged turning away from studies of sources of variation in social ecology. This position might be heuristic because such studies are some of the more difficult and time-consuming that may be carried out (e.g., Anderson and Mason, 1974; Hinde and Spencer-Booth, 1967). However, this position is not tenable if one is interested in evolutionary social ecology.

Identification of sources of variation is necessary for construction of strong evolutionary explanations. In the hypothetical example given above, the importance of differences in habitat selection and homologous learning abilities could only have been unraveled after careful developmental analyses. Thus, there is a mandatory link between evolutionary social ecology and careful, developmental studies of species' differences in social behavior. In retrospect, much of the nature–nurture controversy was concerned with the necessity of the link between developmental studies and evolutionary explanations of differences in the behavior of species (Lehrmann, 1953; Tinbergen, 1969). It would seem unwise to forget this important and difficult lesson.

The above paragraphs have been used to indicate that genetic *differences* must be demonstrated for strong evolutionary explanations of *differences* in social organizations. There is nothing wrong, however, with hypothesizing about selective pressures that may have been responsible for assumed genetic differences. Such hypotheses provide one of the more intellectually stimulating aspects of primate research and, if testable, may guide functional, empirical research on social ecology (e.g., S. Altmann, 1974; Kummer, 1971a; Rasmussen, 1980c). One should, however, keep in mind the chasm separating that which is believed and that which is known.

Many behavioral scientists have been concerned with three related issues: first, the suitability of primate species as subjects in social ecological investigations; second, whether the social ecological questions addressed may be best investigated in natural[6] or in artificial environments; third, whether behavioral adaptive modifications or pathological behaviors may be profitably investigated. The answers to each of these issues are partially determined by whether evolutionary, proximate, or functional social ecological questions are addressed. Therefore, here and at the end of the next two sections of this chapter, these issues are discussed within the context of evolutionary, proximate, and functional social ecology.

Suitability of Primates as Subjects. Comparative and evolutionary analyses of behavior do not require measure of reproductive success or inclusive fitness. Thus, the relatively long lives of primates do not, at first glance, make them more intractable subjects than very short-lived animals, such as many invertebrates. However, analyses of the genetic bases of differences in primate social organizations will be very difficult, because many primates require several years to develop and because of plasticity of their behavior. Thus strong evolutionary reconstructions of the ultimate causes of differences in primate social organizations will be arduous and time-consuming. Primates should therefore only be used as subjects in evolutionary social ecology when they are of intrinsic interest to the investigation. Primates should generally not be used to elucidate approaches to evolutionary problems or to test hypotheses that may be tested on animals with shorter lives. Primates, then, are not the most pragmatic subjects for use in evolutionary social ecology. As discussed in the final section, the greater difficulties in using primates as subjects may have several compensations.

Natural vs. Artificial Environments. In comparative, social ecological analyses current environments of subjects are used often as models of environments believed to have been responsible for selection of differences

---

[6] There are at least two meanings of the term "natural environment." First, a natural environment may refer to the present environment in which a species exists. For example, a large proportion of the rhesus (*Macaca mulatta*) population in northern India lives in urban areas (Southwick *et al.*, 1965). Thus, urban areas form a part of the natural environment in which the rhesus monkey is currently being affected by natural selection. Second, a natural environment may refer to the environment responsible for the evolution of a species (e.g., Bernstein, 1967). This application of the term is unambiguous only when recently diverged species are compared (all primate species have been subjected to many different environments during the course of their entire evolutionary history). For example, if two recently diverged populations are known to differ on some genetic bases, then "natural environments" are environments believed to be identical to those responsible for the genetically based differences between the populations. Here and throughout the rest of this chapter the term natural environment is only used with this second meaning.

observed in social organizations (e.g., Clutton-Brock, 1974). If this approach is used, then it is essential to collect observations in the most natural and undisturbed environments in which the species currently are found. If current environments are used as models of those responsible for the selection of differences in social organizations, then the more recent the evolutionary divergence of the study populations, the more reasonable are the models. This latter statement is true because the greater is the phylogenetic divergence of species, the less likely are the current environments of the species to be good approximations of those responsible for the divergence. For example, comparisons of differences in the social organizations of baboons (*Papio anubis*) and chimpanzees (*Pan troglodytes*) with differences in their current habitats could not be expected to reveal much information about the selective forces responsible for genetic differences in their social organizations. Indeed, chimpanzees and baboons are sympatric in some portions of their present range (Morris and Goodall, 1977; Teleki, 1973).

Social ecologists interested in comparative and evolutionary analyses are confronted with a "double-bind." The more closely related the populations under study, the more likely are the observed differences due to adaptive modifications. The logical extreme here is the comparison of genetically identical groups in different environments. On the other hand, the greater the phylogenetic divergence of study populations, the less will contrasts in their current environments reflect ultimate sources of differences in social organizations (Fig. 2).

Adaptive Modifications vs. Pathological Behavior. Pathological behavior has been defined as behavior elicited by environmental variations believed to extend beyond the range of variation occurring in any natural environment of a given species (Gartlan, 1968). If animals are in "natural" environments, by definition, their behavior may not be termed pathological. Thus, studies on evolutionary social ecology would seem unlikely to be concerned with pathological behavior.

IV. PROXIMATE SOCIAL ECOLOGY

Proximate social ecology is the study of relationships between immediate environmental variation and variation in social organization (Fig. 3). As the terms immediate (Kummer, 1971b) and proximate (Baker, 1938) imply, this subdivision of social ecology is focused on effects of the environment on social organizations during the lifespan of members in those social organizations.

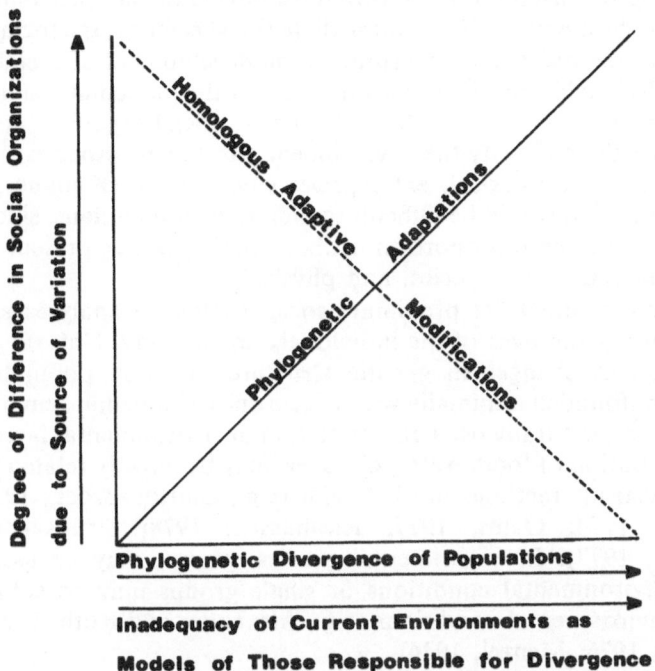

Fig. 2. A diagrammatic representation of a "double-bind" faced in comparative studies of the evolution of differences between social organizations of primate species. The less the phylogenetic divergence between study populations, the more likely are the differences between social organizations due to adaptive modifications. The greater the phylogenetic divergence between populations, the poorer will be their current environments as models of those responsible for the divergence.

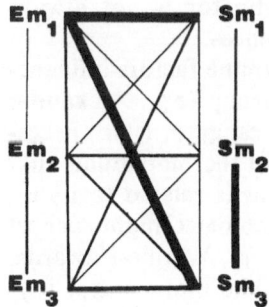

Fig. 3. A simplified and diagrammatic causal network between three measures of the immediate environment (such as distribution of food, sleeping sites, and water) and three measures of aspects of social organization (such as agonistic rank, number of grooming partners, and nearest-neighbor distances to potential mates). The network is not complete, since relationships between variables, as well as the variables themselves, may be of considerable importance (e.g., Hinde and Stevenson-Hinde, 1976). The study of such networks connecting variations in the immediate environment and variation in aspects of social organization is the domain of proximate social ecology.

It may prove useful in the future to subdivide further proximate social ecology into developmental and immediate social ecology, the former focusing effects of the immediate environment on development of social interactions of individuals, and the latter on effects of the immediate environment on more short-term and transitory changes in social organization. It may also be heuristic to classify the environment into the intraspecies, social environment and the physical, extraspecies environment (Kummer, 1971a). This latter division would be difficult to maintain with humans, since we are responsible for a large proportion of the variation in our present environment (intraspecies, extraspecies, and physical).

Data most suited for proximate social ecological analyses are those gathered during the lives of the individuals under study. Natural selection and subsequent changes in genetic structure of study populations are therefore confounded minimally with effects due to immediate environment. In field studies, naturally occurring variation in environmental factors, such as the distribution of food, water, or cover, may be directly related to variations in social interactions and behavior (e.g., Clutton-Brock, 1975; Daly and Daly, 1974; Oates, 1977; Rasmussen, 1978; Struhsaker, 1974; Wrangham, 1977). In experimental studies, groups may be assigned to differing environmental conditions or single groups may be subjected to differing environmental conditions (e.g., Alexander and Roth, 1971; Erwin and Erwin, 1976; Menzel, 1974).

The dependent variables selected for measurement in proximate social ecology are largely determined by the questions addressed. For example, if total levels of agonistic behavior in a group are of interest, then measures of total frequencies and durations of various agonistic patterns of behavior are appropriate. Such studies of total frequencies and durations of behavior among individuals in a social organization may be of use for management of captive groups and for generating models of the effects of the immediate environment on types of behavioral interactions among nonhuman primates and among humans. In proximate social ecology, consequences of environmental variation may be investigated at the level of the individual, group, or population, since the focus of the investigator is not always centered on the adaptive function of those consequences.

For example, an investigator might wish to determine factors influencing the overall level of agonistic behavior within a group (e.g., Alexander and Roth, 1971; Erwin and Erwin, 1976). Such an investigator is not necessarily interested in differences among the behaviors of the individuals that form the group, or in how each individual's behavior is related to its inclusive fitness. An evolutionary or functional social ecologist might suspect such an investigator of advocating group selection (e.g., Wynne-Edwards, 1962); this is not necessarily true. The investigator, like many social psychologists, is interested in group management. The consequence of such group management may well be selection of individuals for certain group characteristics. What is "best" for the group may be least equitable for some individuals (Rasmussen and Rasmussen, 1979). Thus, scientific management of groups may require analyses of behavior at the levels of both the group and the individual.

The multilevel focus of proximate research is one of the most serious impediments to understanding between those interested in primarily evolutionary and functional analyses and those with other motivations for the study of social organizations (see e.g., Blurton Jones, 1976). Investigators whose interests are in evolutionary or functional questions are concerned ultimately with consequences of individuals' social interactions as reflected in changes in gene frequencies in populations (Mayr, 1963; Williams, 1966). Nevertheless, not everyone who studies social organizations is primarily interested in evolutionary or functional questions. For example, those interested in the pressing problems facing humanity engendered by rapid population growth and a drastically altered physical environment (e.g., Calhoun, 1973a,b; Erlich and Erlich, 1972; Quick, 1974; Tinbergen, 1968, 1972) must work with statistical summations reflecting overall levels of interactions. Such summations may be used to describe global effects of immediate environment on social organization and, as illustrated by an example in Section V, provide an important descriptive link for tracing the network connecting environmental variation to the functional consequences of that variation.

Suitability of Primates as Subjects. Animals with long lives are ideally suited for proximate, social ecological analyses. Such animals may be used in several experimental replicates or, in the field, may be observed through several natural seasonal variations in the environment. The longer the lives of the animals, the longer may behavioral scientists study effects of variation in environment confounded minimally by effects of natural selection. Animals with frequent social interactions provide more data for quantitative analyses. Many species of primates are thus ideal for proximate, social ecological analyses.

Natural vs. Artificial Environments. Proximate social ecological analyses may be carried out in either artificial or natural environments. Since the same questions, for example, effects of resource distribution on social behavior, may be investigated in artificial or natural environments, there is no strict dichotomy between analyses carried out in these different environments (Menzel, 1968; Rowell, 1967a; Rumbaugh, 1970; Schneirla, 1950). The immediate environment must, however, be treated as an independent variable because it may exert a profound influence on social organization (e.g., Davis, 1958; Rasmussen and Rasmussen, 1979).

Adaptive Modifications vs. Pathological Behavior.   Adaptive modifications and pathological behavior may both be of interest to the investigator. Indeed, abnormal conditions may be part of an experimental design to investigate normal development (e.g., Struble and Riesen, 1978).

V. FUNCTIONAL SOCIAL ECOLOGY

Functional analyses of social behavior focus on ways in which behavioral interactions among individuals influence their inclusive fitness (Alexander, 1974, 1975; Bertram, 1976; Daly and Wilson, 1978; Hamilton, 1963, 1964; Hinde, 1975; Rasmussen, 1980c; Trivers, 1972; Wilson, 1975). That is, such analyses are concerned with how behavioral interactions among individuals influence the perpetuation of their genes through effects on their offspring and on other members of the population with whom they share genes. Functional social ecology is the study of the ways in which the immediate environment affects those social interactions that influence inclusive fitness. The mandatory use of measures of inclusive fitness, or assumed correlates of inclusive fitness, as dependent variables is thus the differentiation between proximate and functional social ecology.

Functional social ecology may be viewed as the study of the entire chain of social consequences resulting from variation in the proximate environment that affect inclusive fitness. Proximate social ecology is focused on the causal network between environmental and social measures but excludes the relationships between the causal network and measures of inclusive fitness (Fig. 4).

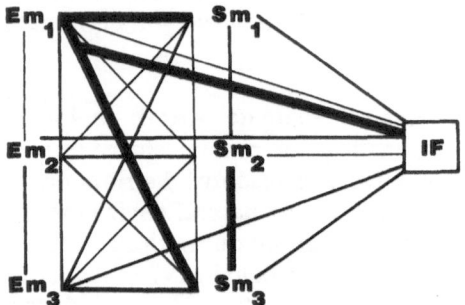

Fig. 4. The same diagrammatic network as in Fig. 3, except that variables in the network as well as relationships between those variables are related to a measure of inclusive fitness. The study of relationships between aspects of such a network and inclusive fitness is the domain of functional social ecology.

The importance of including immediate environmental factors in all functional analyses cannot be overstressed. In general, the environment in which a social organization exists will strongly influence, if not wholly determine, the ratio of the costs to benefits of social interactions. The now classic problem of explaining why one person should save another from drowning (Trivers, 1971) is complicated not only by the degree of relatedness between the two parties and by chances of reciprocity, but also by their environment. Obviously, if a person is drowning in a wading pool the costs associated with saving that person will be considerably less than if the person is drowning in a stormy surf. Further, the cost/benefit ratio and the classification of this "altruistic" action is radically changed if we find the "altruist" is exploiting the "niche" of a lifeguard. If we find the "altruist" is a lifeguard, then saving another person from drowning is simply an indirect method of harvesting resources (see also Allee, 1943, and West-Eberhard, 1975). We may find "altruists" tend to win popularity, resources, and greater access to mates. Taking "altruistic" risks may thus be a sexually selected activity (Zahavi, 1975). Indeed, "altruistic" behavior may function as a mark of quality and be a product of intrasexual selection (O'Donald, 1962, 1963).

There are a multitude of potential research designs that may be applied in functional social ecology. Patterson (1975) has suggested a simple and appealing design suited for field research. Naturally occurring environmental variation can be measured and related to measured variations in social behavior, and those variations in social behavior can then be related to a measure of inclusive fitness. For example, adult males in a study population might be observed to form a rigid agonistic rank-order at certain times of the year whereas an overlapping agonistic rank-order might be observed at other times (Fig. 5). Let us suppose that in the former case, males' positions in the hierarchy are more strongly and postively correlated with a measure of inclusive fitness than in the latter. From such a study, it could be concluded that the difference in environmental conditions responsible for the more linear rank-order resulted in stronger selection for heritable characteristics associated with high agonistic rank in adult males. One might hypothesize that a permanent change in the environment to those conditions associated with the more linear agonistic rank-order would result in an increase in future populations of those genes associated with high agonistic rank in adult males.

The delimitation, emphasis, and repeated underscoring of measures of inclusive fitness as the ultimate biological dependent variables may well be the single greatest contribution of sociobiology to functional social ecology. Perhaps this point is best illustrated by a return to Crook's (1970b) use of a football game as an analogy of a social organization.

The problem is for a behavioral scientist, completely ignorant of football, to empirically describe the game, understand its rules, and discover its

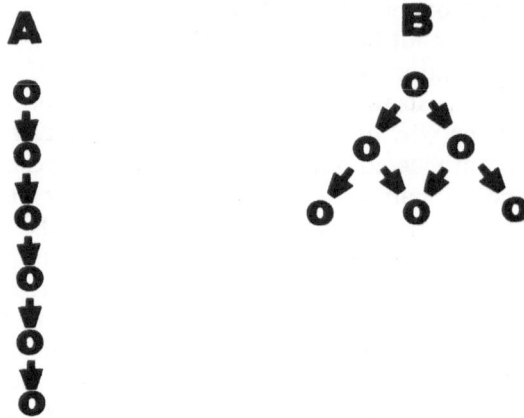

Fig. 5. *A* is a linear agonistic rank-order. *B* is an overlapping agonistic rank-order (after Dawkins, 1976). It might be found that at certain times of year a linear rank-order was formed by adult males, whereas at other times of year an overlapping agonistic rank-order was present. If a stronger positive correlation between rank and a measure of inclusive fitness exists in the former case, then one might hypothesize that a permanent change to the environmental factors responsible for the more linear rank-order would result in an increase in the population of those genes associated with high agonistic rank in adult males.

function. Dyadic analyses of all social interactions are a rather uninspired means of gaining an understanding of the game. Dyadic analyses may be useful for elucidating aspects of the game; however, such analyses seem unlikely to be adequate for the task of describing the entire game (Anderson and Mason, 1978; Kummer, 1967a). Many have suggested the study of roles as an alternative and better method (Bernstein, 1964, 1966, 1970; Bernstein and Sharpe, 1966; Crook, 1970b; Gartlan, 1968; Rowell, 1966). Focus on roles is a considerable conceptual advance over focus on dyadic interactions. However, even with the use of multivariate methods to quantitatively describe roles (Fairbanks *et al.*, 1978; Fedigan, 1976), choice of behavioral categories (Bateson, 1976; Dunbar, 1976; Menzel, 1979), sampling procedures (J. Altmann, 1974), and role definitions are somewhat arbitrary and the success of the analysis of our football game will be to a large extent contingent upon how well the investigator already understands the game when the behavioral categories and sampling procedures are devised.[7] Spatial analyses of primate social organizations seem likely eventually to

[7] I doubt if even the most sophisticated quantification of behavioral analyses will ever completely replace the importance of simply watching social interactions in order to first obtain an "intuitive" sensitivity to the relative importance of social interactions.

provide the most quantitative and exact descriptions of those organizations (Brown and Orians, 1970; Carpenter, 1964; Kummer, 1971c; Menzel, 1968, 1979; Rasmussen, 1980b). Spatial analyses are perhaps best used to describe a social organization as it exists, and not to describe the way in which it is maintained or formed by interactions among individuals (Rasmussen, 1973).

Fortunately the behavioral scientist is not ignorant about the function of social organizations. The function, or the reason why individuals form social organizations, is for each individual to increase its inclusive fitness (e.g., Alexander, 1974, 1975; Dawkins, 1978; Hamilton, 1963, 1964; Wilson, 1975).[8] The power of the knowledge of this function has been only recently realized and holds incredible potential for all behavioral research. To those who think in terms of statistical analyses, this knowledge allows a shift from multivariate methods, such as factor analysis, that organize data into groups of variables, to multivariate methods, such as multiple regression and multiple correlation, that relate multiple variables and relationships between variables to a single dependent variable. This shift is accompanied by a considerable reduction in arbitrary statistical assumptions. Measures of inclusive fitness may provide the sorely needed standards by which other behavioral variables can be evaluated.

To return to our football game: if players are not relatives, then it is possible to let their salaries be rough analogues to their inclusive fitness.[9] Knowledge that salaries of players are the key to the functional analysis of their behavior provides a focus for the analysis of the game. Use of salaries as the dependent variable permits evaluation of the relative importance of various aspects of behavior and measures of organization of the game. For example, each player's running speed, the speed with which the ball is passed between dyads of players, roles of the players, membership in cliques within each team (Sade, 1972), the relative frequency with which one team scores points relative to others, spatial relationships among group members, and number of games won by each team can all be evaluated and related to salaries of members of each team. Immediate environmental factors that might be strongly related to salaries of players are number of spectators at each game, number of TV cameras at each game, and quality of the football

---

[8] Given that the paradigm of natural selection is accepted to be true, this is a tautologous statement: The genetic substrates of the behavior of individuals whose social interactions increase their inclusive fitness must increase in all populations. Nevertheless, recognition of the logically irrefutable nature of this statement represents a major advance in the conceptual analysis of social organizations.

[9] This analogy is not exact since inclusive fitness "won" in social interactions may only be "spent" in future games. Salaries, of course, may be used elsewhere (Slobodkin, 1978).

field on which games are played. This type of analysis would not only indicate the relative importance of each variable, but would also indicate, empirically, the "best" way to measure each variable [the "best" being the measure which most strongly and reliably predicts the players' salaries (see Anastasi, 1968; Cohen and Cohen, 1975; and Petrinovich, 1981, for further elaboration of this viewpoint)].

I would like to move away from this analogy now and compare two primate societies with which I have had first-hand experience. This less abstract example may further illustrate the potential of the approach advocated here.

The first society I studied was the troop of Japanese macaques (*Macaca fuscata*) housed at the Oregon Regional Primate Research Center; I observed this troop for about 400 hours. In this troop there seems to be no correlation between the agonistic rank of the adult males and copulation frequency (Eaton, 1974). Arrowhead, the male of highest agonistic rank, has maintained his position since the troop was first brought to Oregon in 1964 (Eaton, 1976, and personal communication). Yet Arrowhead has no canines, has lost one eye, and is considerably smaller than many of the other males. Subjectively, Arrowhead and other males of old age look as if they are "respected" by fellow troop members.

The second society I studied was Viramba troop, a troop of about 120 yellow baboons (*Papio cynocephalus*) located in Mikumi National Park, Tanzania; I spent over 4000 hours within 15 meters of the members of this troop. As will be indicated in future publications, there was a correlation between the agonistic rank of the males and the frequency with which they copulated with impregnable females (see also Hausfater, 1975). In this troop, males started to decrease in agonistic rank as soon as their body condition appeared slightly short of perfect and their canines even a little less than needle sharp (see also Packer, 1977, 1979). To me, the life of an old male in Viramba troop looked very unpleasant indeed.

Perhaps the apparently very different fates of old males is a consequence of the ways in which their behavior may continue to contribute to their inclusive fitness. For example, the male composition of the Oregon troop is relatively static, since emigrations and immigrations by adult males cannot naturally occur in a confined troop. Over 81% of the 32 adult males who were observed within Viramba troop emigrated from or immigrated into that troop during the 22 months that daily censuses were collected. It seems rather likely that the old adult male Japanese macaques would tend to have more offspring within their troop than the old adult male baboons in Viramba troop. Consequently, a greater proportion of energy devoted to protection and care of troop members, rather than to copulation and intrasexual competition, might more strongly increase the inclusive fitness of the old adult male Japanese macaques than the baboons (e.g., Allee, 1943; Eshel, 1972; Wilson, 1975; Wright, 1945). Such differences in the relative values of activities might be causally related to the presence or absence of a correlation between agonistic rank and copulation frequency in these two different primate societies. Such speculations are imminently open to empirical analysis.

Functional social ecology is still in an embryonic state. One major obstacle standing in the way of its development is the difficulty in obtaining measures of inclusive fitness (Bertram, 1976; Trivers, 1971). Certainly, measures of individuals' reproductive success (e.g., Lack, 1954), or a probable correlate of reproductive success (e.g., Bernstein, 1976; Hausfater, 1975), serve as first approximations. Yet differences between individuals' reproductive success and their inclusive fitness may well be crucial information for understanding differences observed between social organizations (Hamilton, 1964).

Determination of an individual's influence on its inclusive fitness is even more difficult than measurement of inclusive fitness. Two parameters must be determined: first, how the assortment of genes carried by an individual is propagated when that individual lives out its life; second, how those genes are propagated if that individual never existed. Subtraction of the value of the second parameter from that of the first yields the influences of the individual on its inclusive fitness. How can one measure the effects of the existence of a unique diploid primate on the propagation of the genes it carries and shares in common with other individuals? What, for example, would the present genetic structure of humanity be like if Moses, Christ, Mohammed, Darwin, and Hitler had not existed?

Perhaps I paint an overly grim picture of the problems facing those who wish eventually to measure inclusive fitness. It seems likely that several different measures will be devised, all less than perfect. For example, the effects of the presence or absence of postreproductive females on the social organization of a given primate species could be measured by comparing the offspring produced in matched groups, some with such females present, some with them absent. Results from such an experiment would give an approximate answer to how the presence of such females affected offspring produced in the groups.

Another difficulty with measures of inclusive fitness and reproductive success is that they may only be given a value relative to that of other individuals at a given point in time. As we observe a social organization, we tend to assume that individuals with the highest inclusive fitness during the period of study are those who are most successful. Suppose, however, that a short time after the termination of a study, environmental conditions change or there was an unpredicted change in social organization. Such changes could drastically alter evaluations of inclusive fitness. Just as historians must continually rewrite history in the face of changes after events, so too will behavioral scientists be forced to alter evaluations of inclusive fitness and reproductive success after the passage of time. As economists and sociologists have long been aware, extrapolation of data in time is always a hazardous and an inexact procedure (e.g., Johnston, 1972; Spilerman, 1975).

Suitability of Primates as Subjects. Ideal animals for close study in functional social ecological analyses are those with very short intervals between generations, such as *Drosophila*. In such animals, effects of environmental conditions on social interactions may be measured and related to reproductive success or a measure of inclusive fitness; resultant changes in the genotypic structure of the population may be assessed. Unfortunately, primates are generally very poor subjects for functional analyses.

Why then should primates ever be used as subjects in functional analyses? Perhaps the most general reason is that their behavior is so extremely flexible to proximate environmental variation. Functional analyses of primate behavior will therefore provide a necessary counterweight to functional analyses carried out on organisms whose almost only means of adapting to environmental changes is genetic.

Those primates most suited for functional social ecological analyses are those in which: (1) frequent social interactions occur, (2) large disparities in reproductive success exist, and (3) some individuals leave many offspring during a relatively short time period. Such conditions are found in males who live in large polygynous groups such as yellow baboons, *Papio cynocephalus* (Hall and DeVore, 1965; Rasmussen, 1980b). In females, there is less likely to be as large a range in reproductive success (Trivers, 1972). As a result, future empirical analyses of the functional social ecology of primates seem likely to have a bias in favor of polygynous males.

If only one generation of offspring is studied, it will be possible to estimate only how the immediate environment is currently affecting the selection of individuals. Nothing may be said about the direction in which the population is being selected, as selection pressures may have been stronger or weaker in preceding generations.

Natural vs. Artificial Environments. As with proximate social ecology, functional social ecology does not have to be limited to "natural" environments. Functional social ecology focuses on the influences of the immediate environment on inclusive fitness and thus may be carried out on animals in an unnatural or an artificial environment. For example, examination of possible influences of the behavior of individuals in captivity on their fitness

(Modahl and Eaton, 1977) may suggest behavioral traits likely to be selected in captive groups and how these traits may differ from those selected in natural groups. Results from such studies will be important in the future when many of the behavioral analyses of primates will be carried out on subjects that have been reared for several generations in captive colonies. Such studies may also yield models predicting effects of drastic environmental changes on social organization (e.g., Davis, 1958), and yield information on how selection pressures might change if a given species invaded a new habitat.

Adaptive Modifications vs. Pathological Behavior.   There is no reason why interest should be limited to adaptive modifications in functional social ecology. Behavior patterns are evaluated in terms of inclusive fitness and not in terms of "normality." Indeed, pathological behavioral patterns may be found to be strongly correlated with major differences in individuals' inclusive fitness; if so, this information may be of importance to future generations of researchers working with "domesticated" primate stock.

VI. INTERDIGITATION OF EVOLUTIONARY, PROXIMATE, AND FUNCTIONAL SOCIAL ECOLOGY

The tripartite division of social ecology discussed in the preceding three sections is a conceptual division. In practice, many who do research in social ecology use an exceptionally complex blend of all three approaches as did Tinbergen (1965) in his discussion of behavior and natural selection. In this section, a few of the more obvious strands of connection between evolutionary, proximate, and functional social ecology are discussed.

Comparative data from primate social ecology used in evolutionary reconstructions contain variation due both to the phylogenetic history of the study populations and to their immediate environment. Nevertheless, judicious use of such confounded data may lead to useful empirical generalities. For example, several authors have attempted to relate attributes of the habitats in which primate species are found with aspects of their social organizations (Crook and Gartlan, 1966; Denham, 1971; Eisenberg et al., 1972; Jolly, 1972). Although these early attempts met with only limited success (S. Altmann, 1974; Clutton-Brock, 1974; Struhsaker, 1969; Wilson, 1975), they did center attention on relationships between environment and social organization and to the possibility of a scientific understanding of the interrelations between social organization and ecology.

In contrast to efforts to relate attributes of habitats to aspects of primate social organization, an emphasis on the way habitats are used seems likely to lead to more valid empirical generalities. The immediate habitat in which a study population is observed may be very different from that responsible for genetic differences between its social organization and the social organizations of other populations. However, the way the study population uses that habitat will reflect both its phylogeny (e.g., dietary and morphological constraints) and the influence of the immediate habitat (e.g., the distribution and abundance of food patches). Similarly, the social organization of the study population will reflect its phylogeny and its immediate environment (e.g., Kummer, 1971a; Struhsaker, 1969; Wilson, 1975). Thus, correlation of aspects of use of habitats with aspects of social organization seem, in theory, more likely to reveal important biological relationships since both sets of variables must contain variation due both to the phylogeny of the subjects and to the immediate environment.

Recent comparisons of primate ecology and social organization have met with greater success than earlier attempts (Clutton-Brock and Harvey, 1977; Jorde and Spuhler, 1974; Milton and May, 1976). This greater success is probably partially due to use of measures, such as day range length and home range size, as "ecological factors." These factors are, of course, measures of *use* of the environment. Such studies using confounded sources of variation do not indicate differences in selective pressures or evolutionary trends. Rather, they indicate adaptive aspects of individuals' social behavior, aspects due to an unknown blend or phylogenetic adaptations and adaptive modifications.

Proximate studies form nearly the entire data base of primate social ecology (e.g., the studies contained in the volume edited by Clutton-Brock, 1977). A large number of these studies have been concerned with the manner in which primates use their environment. Such studies indicate: (1) how animals' movements are controlled by an incredible diversity of "resources" and "hazards"[10] that occur in their environment (e.g., Altmann and Altmann, 1970; Chivers, 1974; Clutton-Brock, 1975; Daly and Daly, 1974; McKey et al., 1978; Oates et al., 1977; Rodman, 1973); (2) the nature of those "resources" and "hazards" (e.g., Chivers, 1974; Hladik, 1975); and (3) the way in which those "resources" and "hazards" affect, and are related to, social interactions (Calhoun, 1971; Rasmussen, 1979; Rasmussen and Rasmussen, 1979; Struhsaker, 1974; Wrangham, 1977).

---

[10] Jander (1975) uses the term "stress sources." "Hazards" seems to be a more general term. Resources are all those factors which increase the inclusive fitness of an individual. Hazards are those factors which decrease the inclusive fitness of an individual. In general, resources should elicit approach; hazards should elicit withdrawal (e.g., Schneirla, 1965), although this is not always strictly true (Kruuk, 1976).

Oddly, few primate social ecologists directly measure effects of environment, or use of environment, on social interactions. Most investigate the nature of habitats of primates or primates' use of their habitats. The "social" of primate social ecology has therefore been largely left out. This neglect of direct measurement of effects of immediate environment on social interactions may be due in part to the extreme importance of evolutionary hypotheses for giving impetus to the study of primate social ecology (e.g., Washburn and DeVore, 1961). As a result, researchers have tended to design studies for examination of evolutionary aspects of primate social organizations. For example, Clutton-Brock (1974) discusses how *differences* in ways red (*Colobus badius*) and black and white (*C. guereza*) colobus use food sources may have selected for *differences* in their social organizations.

If the minds of primate social ecologists have been guided by evolutionary considerations, their feet have been firmly and directly planted in proximate social ecology. It seems to me that proximate social ecology in and of itself has many important applications, in addition to providing essential data for evolutionary reconstructions. Perhaps more direct recognition of the proximate nature of most data would result in research reports of even greater interest to the general scientific community. Certainly, exact and quantitative analyses of ways in which distribution of resources and use of those resources affect behavior may find immediate application in management of captive animals (Hediger, 1964; Markowitz and Stevens, 1978; Rasmussen and Rasmussen, 1979; Reinhardt et al., 1978; Wilson, 1979). Analyses of those patterns of social interactions that are found to be affected by the conditions of captivity may also be used in a manner roughly analogous to the "deprivation experiment" (Lorenz, 1965). Interaction patterns that remain similar to those observed in wild populations are less labile and thus singled out as possibly rigidly controlled through processes such as genetic canalization (Waddington, 1957) or through nongenetic transmission of information across generations (Kummer, 1967b, 1971a; Kummer and Kurt, 1965; Rowell, 1967b). Such captive-wild comparisons, like deprivation experiments (Bateson, 1976), may be useful precursors to more careful developmental analyses.

Studies of effects of drastically altered environments on primates with different modal social organizations (e.g., Hall, 1963; Kummer, 1971a) may also reveal empirical generalities (e.g., Collias, 1944) about effects of environmental manipulations on primates with different modal social organizations. Finally, studies of effects of systematic physical and social environmental manipulation on social interactions of primates (Alexander and Roth, 1971; Marsden, 1972; Mason, 1968, 1971; Southwick et al., 1976) eventually may yield models and hypotheses about effects of further environmental changes on human social organization.

Functional social ecology is potentially a rigorous empirical science. It is separated from proximate social ecology by dependent variables that measure inclusive fitness. Data from functional social ecology provide information on how proximate environmental variation is currently affecting the selection of individuals with different phenotypes. Naturally, such data do not indicate genetic bases of the differences in phenotypes, one of the important domains of behavior-genetic research (Hirsch, 1967; Manning, 1975). If functional social ecological studies are conducted in natural environments, then the data provide one of the strongest possible bases for reconstruction of how past forces of natural selection may have led to presumed genetic differences (Bertram, 1976).

Sociobiological theory seems likely eventually to provide the necessary framework by which studies of primate behavior and primate social ecology may be ordered (e.g., Clutton-Brock and Harvey, 1976). The theory is particularly welcome when it leads to counterintuitive hypotheses, hypotheses that might not otherwise have been tested by an inductive scientist embedded in his or her own cultural and valuational system. The work of Trivers (1971, 1972, 1974) has been notable for generating such hypotheses and counter-hypotheses. The theory is also helpful to the empiricist when it provides conceptual tools such as the game theories developed by Maynard Smith (e.g., 1976).

Little is actually known about the functional or evolutionary social ecology of any primate species. It is possible to use evolutionary paradigms derived from analyses of more readily studied species such as *Drosophila* to "explain" variation in primate social organizations. Such use of evolutionary paradigms is exactly analogous to the way behaviorists have used learning paradigms derived from studies of laboratory animals such as white rats (Beach, 1950) to "explain" most of human behavior. Heavy and uncritical reliance on such evolutionary paradigms therefore should be equally suspect. As Hinde and Stevenson-Hinde (1976, p. 462) have remarked in another context: "If one uses a paradigm at more than a descriptive level, or with too heavy a hand, then its advantages are lost."

It is at the point of exacting analyses of single species that inconsistencies may be found in broad sweeps of theories and actual contributions of phylogenetic and adaptive modifications may be discerned. At the level of nonhuman primates, functional and evolutionary analyses of social orgnizations will meet with the need for greatest exactitude and least reliance on descriptive paradigms for filling in the gaps in our knowledge. This is true both because of the incredible plasticity of primate behavior and because when we study nonhuman primates we are but one step removed from study of ourselves.

VII. ACKNOWLEDGMENTS

I thank Dr. P. P. G. Bateson for his insightful and scholarly comments on the penultimate draft of this chapter, Dr. W. A. Mason for helping me to clarify several passages in the text, Drs. J. A. Moore, L. F. Petrinovich, and A. H. Riesen for comments on manuscripts I wrote in 1972 and 1973 that touched on some of the topics in the section on evolutionary social ecology, and Mr. D. D. Cubicciotti III and Mr. R. W. Summers for their useful comments and careful reading of this chapter. The production of this paper was partially supported by National Institutes of Health Grant No. RR00169.

VIII. REFERENCES

Alexander, B. K., and Roth, E. M. (1971). The effects of acute crowding on aggressive behavior of Japanese monkeys. *Behavior* **39**:73—90.
Alexander, R. D. (1974). The evolution of social behavior. *Annu. Rev. Ecol. Syst.* **5**:325—383.
Alexander, R. D. (1975). The search for a general theory of behavior. *Behav. Sci.* **20**:77—100.
Allee, W. C. (1943). Where angels fear to tread: a contribution from general sociology to human ethics. *Science* **97**:517—525.
Altmann, J. (1974). Observational study of behavior: sampling methods. *Behavior* **49**:227—267.
Altmann, S. A. (1974). Baboons, space, time, and energy. *Am. Zool.* **14**:221—248.
Altmann, S. A., and Altmann, J. (1970). *Baboon Ecology: African Field Research*, University of Chicago Press, Chicago.
Anastasi, A. (1968). *Psychological Testing*, 3rd ed., Macmillan, London.
Anderson, C. O., and Mason, W. A. (1974). Early experience and complexity of social organization in groups of young rhesus monkeys (*Macaca mulatta*). *J. Comp. Physiol. Psychol.* **87**:681—690.
Anderson, C. O., and Mason, W. A. (1978). Competitive social strategies in groups of deprived and experienced rhesus monkeys. *Devel. Psychobiol.* **11**:289—299.
Baker, J. R. (1938). The evolution of breeding seasons. In: de Beer, G. R. (ed.), *Evolution*, Oxford University Press, Oxford, pp. 161—171.
Bateson, P. P. G. (1976). Specificity and the origins of behavior. *Adv. Stud. Behav.* **6**:1—20.
Beach, F. A. (1950). The snark was a Boojum. *Am. Psychol.* **5**:115—124.
Bernstein, I. S. (1964). Role of the dominant male rhesus monkey in response to external challenges to the group. *J. Comp. Physiol. Psychol.* **57**:404—406.
Bernstein, I. S. (1966). Analysis of a key role in a Capuchin (*Cebus albifrons*) group. *Tulane Stud. Zool.* **13**:49—54.
Bernstein, I. S. (1967). Defining the natural habitat. In: Starck, D., Schneider, R., and Kuhn, H. (eds.), *Progress in Primatology*, Gustav Fischer, Stuttgart, pp. 177—179.
Bernstein, I. S. (1970). Primate status hierarchies. In: Rosenblum, L. A. (ed.), *Primate Behavior*, Vol. 1, Academic Press, New York, pp. 71—109.
Bernstein, I. S. (1976). Dominance, aggression, and reproduction in primate societies. *J. Theor. Biol.* **60**:459—472.

Bernstein, I. S., and Sharpe, L. G. (1966). Social roles in a rhesus monkey group. *Behavior* **26**:91–104.

Bertram, B. C. R. (1976). Kin selection in lions and evolution. In: Bateson, P. P. G., and Hinde, R. A. (eds.), *Growing Points in Ethology*, Cambridge University Press, Cambridge, pp. 281–301.

Blurton Jones, N. L. (1976). Growing points in human ethology: another link between ethology and the social sciences? In: Bateson, P. P. G., and Hinde, R. A. (eds.), *Growing Points in Ethology*, Cambridge University Press, Cambridge, pp. 427–450.

Boehm, C. (1978). Rational preselection from hamadryas to *Homo sapiens*: the place of decisions in adaptive process. *Am. Anthropol.* **80**:265–296.

Brown, J. L., and Orians, L. H. (1970). Spacing patterns in mobile animals. *Annu. Rev. Ecol. Syst.* **1**:239–262.

Calhoun, J. B. (1971). Space and the strategy of life. In: Esser, A. H. (ed.), *Behavior and Environment*, Plenum, New York, pp. 329–387.

Calhoun, J. B. (1973a). Death squared: the explosive growth and demise of a mouse population. *Proc. R. Soc. Med.* **66**:80–88.

Calhoun, J. B. (1973b). From mice to men. *Trans. Stud. Col. Phys. Philadelphia* **41**:92–118.

Carpenter, C. R. (1964). *Naturalistic Behavior of Nonhuman Primates*, Pennsylvania State University Press, University Park.

Chivers, D. J. (1974). The siamang in Malaysia. *Contrib. Primatol.* **4**:1–335.

Clutton-Brock, T. H. (1974). Primate social organization and ecology. *Nature (Lond.)* **250**:539–542.

Clutton-Brock, T. H. (1975). Ranging behavior of red colobus (*Colobus badius tephrosceles*) in the Gombe National Park. *Anim. Behav.* **23**:706–722.

Clutton-Brock, T. H. (ed.) (1977). *Primate Ecology: Studies of Feeding and Ranging Behavior in Lemurs, Monkeys, and Apes*, Academic, London.

Clutton-Brock, T. H., and Harvey, P. H. (1976). Evolutionary rules and primate societies. In: Bateson, P. P. G., and Hinde, R. A. (eds.), *Growing Points in Ethology*, Cambridge University Press, Cambridge, pp. 195–237.

Clutton-Brock, T. H., and Harvey, T. H. (1977). Primate ecology and social organization. *J. Zool. Lond.* **183**:1–39.

Cohen, J., and Cohen, P. (1975). *Applied Multiple Regression/Correlation Analysis for the Behavioral Sciences*, Lawrence Erlbaum Assoc., Hillsdale, N.J.

Collias, N. E. (1944). Aggressive behavior among vertebrate animals. *Phys. Zool.* **17**:83–123.

Crook, J. H. (1970a). The socio-ecology of primates. In: Crook, J. H. (ed.), *Social Behavior of Birds and Mammals*, Academic, London, pp. 103–166.

Crook, J. H. (1970b). Social organization and the environment: aspects of contemporary social ethology. *Anim. Behav.* **18**:197–209.

Crook, J. H., and Gartlan, J. S. (1966). Evolution of primate societies. *Nature* **210**:1200–1203.

Daly, M., and Daly, S. (1974). Spatial distribution of a leaf-eating Saharan gerbil (*Psammomys obesus*) in relation to its food. *Mammalia* **38**:591–603.

Daly, M., and Wilson, M. (1978). *Sex, Evolution, and Behavior*, Duxbury, North Scituate, Massachusetts.

Davis, D. E. (1958). The role of density in aggressive behavior of house mice. *Anim. Behav.* **6**:207–210.

Dawkins, R. (1976). Hierarchical organization: a candidate principle for ethology. In: Bateson, P. P. G., and Hinde, R. A. (eds.), *Growing Points in Ethology*, Cambridge University Press, Cambridge, pp. 7–54.

Dawkins, R. (1978). *The Selfish Gene*, Paladin, London.

Denham, W. W. (1971). Energy relations and some basic properties of primate social organization. *Am. Anthropol.* **73**:77–95.

Dunbar, R. I. M. (1976). Some aspects of research design and their implications in the observational study of behavior. *Behavior* **58**:78—98.

Eaton, G. G. (1974). Male dominance and aggression in Japanese macaque reproduction. In: Montagna, W., and Sadler, W. A. (eds.), *Reproductive Behavior*, Plenum, New York, pp. 287—297.

Eaton, G. G. (1976). The social order of Japanese macaques. *Sci. Am.* **235**:96—106.

Eisenberg, J. F., Muckenhirn, N. A., and Rudran, R. (1972). The relation between ecology and social structure in primates. *Science* **176**:863—874.

Erlich, P. R., and Erlich, A. H. (1972). *Population, Resources, Environment: Issues in Human Ecology*, W. H. Freeman, San Francisco.

Erwin, N., and Erwin, J. (1976). Social density and aggression in captive groups of pigtail monkeys. *Appl. Anim. Ethol.* **2**:265—269.

Eshel, I. (1972). On the neighbor effect and the evolution of altruistic traits. *Theor. Pop. Biol.* **3**:258—277.

Fairbanks, L. A., McGuire, M. T., and Page, N. (1978). Social roles in captive vervet monkeys (*Cercopithecus aethiops sabaeus*). *Behav. Processes* **3**:335—352.

Fedigan, L. M. (1976). A study of roles in the Arashiyama West troop of Japanese monkeys (*Macaca fuscata*). *Contrib. Primatol.* **9**:1—95.

Garcia, J., McGowan, B. K., and Green, K. F. (1972). Sensory quality and integration: constraints on conditioning. In: Black, A. H., and Prokasy, W. F. (eds.), *Classical Conditioning. Vol. 2*. Appleton-Century-Crofts, New York, pp. 3—27.

Gartlan, J. S. (1968). Structure and function in primate society. *Folia Primatol.* **8**:89—120.

Hall, K. R. L. (1963). Variations in the ecology of the Chacma baboon (*Papio ursinus*). *Symp. Zool. Soc. Lond.* **10**:1—28.

Hall, K. R. L., and DeVore, I. (1965). Baboon social behavior. In: DeVore, I. (ed.), *Primate Behavior: Field Studies of Monkeys and Apes*, Holt, Rinehart, & Winston, New York, pp. 53—110.

Hamilton, W. D. (1963). The evolution of altruistic behavior. *Am. Zool.* **97**:354—356.

Hamilton, W. D. (1964). The genetical evolution of social behavior I, II. *J. Theor. Biol.* **7**:1—52.

Hausfater, G. (1975). Dominance and reproduction in baboons (*Papio cynocephalus*). *Contrib. Primatol.* **7**:1—150.

Hediger, H. (1964). *Wild Animals in Captivity*, Dover, New York.

Hinde, R. A. (1970). *Animal Behavior: A Synthesis of Ethology and Comparative Psychology*, 2nd ed., McGraw-Hill, New York.

Hinde, R. A. (1975). The concept of function. In: Baerends, G., Beer, C., and Manning, A. (eds.), *Function and Evolution in Behavior*, Clarendon, Oxford, pp. 3—15.

Hinde, R. A., and Spencer-Booth, Y. (1967). The behavior of socially living rhesus monkeys in their first two and a half years. *Anim. Behav.* **15**:169—196.

Hinde, R. A., and Stevenson-Hinde, J. (eds.) (1973). *Constraints on Learning: Limitations and Predispositions*, Academic, London.

Hinde, R. A., and Stevenson-Hinde, J. (1976), Towards understanding relationships: dynamic stability. In: Bateson, P. P. G., and Hinde, R. A. (eds.), *Growing Points in Ethology*, Cambridge University Press, Cambridge. pp. 451—479.

Hinde, R. A., and Tinbergen, N. (1958). The comparative study of species-specific behavior. In: Roe, A., and Simpson, G. G. (eds.), *Behavior and Evolution*, Yale University Press, New Haven, pp. 251—268.

Hirsch, J. (ed.) (1967). *Behavior-Genetic Analysis*, McGraw-Hill, New York.

Hladik, C. M. (1975). Ecology, diet, and social patterning of old and new world primates. In: Tuttle, R. H. (ed.), *Socioecology and Psychology of Primates*, Mouton, The Hague, pp. 3—36.

Jander, R. (1975). Ecological aspects of spatial orientation. *Annu. Rev. Ecol. Syst.* **6**:171—188.

Johnston, J. (1972). *Econometric Methods*, 2nd ed., McGraw-Hill, New York.

Jolly, A. (1972). *The Evolution of Primate Behavior*, Macmillan, New York.

Jorde, L. B., and Spuhler, J. N. (1974). A statistical analysis of selected aspects of primate demography, ecology, and social behavior. *J. Anthropol. Res.* **30**:149–224.

Klopfer, O. H., and Hailman, J. P. (1965). Habitat selection in birds. *Adv. Stud. Behav.* **1**: 279–303.

Kruuk, H. (1976). The biological function of gull's attraction towards predators. *Anim. Behav.* **24**:146–153.

Kummer, H. (1967a). Tripartite relations in hamadryas baboons. In: Altmann, S. A. (ed.), *Social Communication among Primates*, University of Chicago Press, Chicago, pp. 63–71.

Kummer, H. (1967b). Dimensions of a comparative biology of primate groups. *Am. J. Phys. Anthropol.* **27**:357–366.

Kummer, H. (1971a). *Primate Societies: Group Techniques of Ecological Adaptation*, Aldine-Atherton, Chicago.

Kummer H. (1971b). Dimensions of a comparative biology of primate groups. *Am. J. Phys. Primatol.* **3**:1–11.

Kummer, H. (1971c). Spacing mechanisms in social behavior. In: Eisenberg, J. F., and Disson, W. S. (eds.), *Man and Beast: Comparative Social Behavior*, Smithsonian, Washington, D.C., pp. 221–234.

Kummer, H., and Kurt, F. (1965). A comparison of social behavior in captive and wild hamadryas baboons. In: Vagtborg, H. (ed.), *The Baboon in Medical Research*, University of Texas Press, Austin, pp. 65–80.

Lack, D. (1954). *The Natural Regulation of Animal Numbers*, Oxford University Press, London.

Lehrmann, D. S. (1953). A critique of Konrad Lorenz' theory of instinctive behavior. *Q. Rev. Biol.* **28**:337–363.

Lorenz, K. (1950). The comparative method in studying innate behavior patterns. *Symp. Soc. Exp. Biol.* **4**:221–268.

Lorenz, K. (1965). *Evolution and Modification of Behavior*, University of Chicago Press, Chicago.

Manning, A. (1975). Behavior genetics and the study of behavioral evolution. In: Baerends, F., Beer, C., and Manning, A. (eds.), *Function and Evolution in Behavior*, Clarendon, Oxford.

Markowitz, H., and Stevens, V. J. (eds.) (1978). *Behavior of Captive Wild Animals*, Nelson-Hall, Chicago.

Marsden, H. M. (1972). The effect of food deprivation on intergroup relations in rhesus monkeys. *Behav. Biol.* **7**:369–374.

Mason, W. A. (1968). Early social deprivation in nonhuman primates: implications for human behavior. In: Glass, D. (ed.), *Environmental Influences*, Rockefeller University Press, New York, pp. 70–101.

Mason, W. A. (1971). Field and laboratory studies of social organization in *Saimiri* and *Callicebus*. In: Rosenblum, L. A. (ed.), *Primate Behavior*, Vol. 2, Academic, New York, pp. 107–137.

Maynard Smith, J. (1976). Evolution and the theory of games. *Am. Sci.* **64**:41–45.

Mayr, E. (1963). *Animal Species and Evolution*, Belknap, Cambridge, Mass.

McKey, D., Waterman, P. G., Mbi, C. N., Gartlan, J. S., and Struhsaker, T. T. (1978). Phenolic content of vegetation in two African rain forests: ecological implications. *Science* **202**:61–64.

Menzel, E. W. (1968). Naturalistic and experimental approaches to primate behavior. In: Willems, E., and Raush, H. (eds.), *Naturalistic Viewpoints in Psychology*, Holt, Rinehart, & Winston, New York, pp. 78–121.

Menzel, E. W., Jr. (1974). A group of young chimpanzees in a one-acre field. In: Schrier, A. M., and Stollnitz, F. (eds.), *Behavior of Nonhuman Primates: Modern Research Trends*, Vol. 5, Academic, New York, pp. 83—153.

Menzel, E. W. (1979). General discussion of the methodological problems involved in the study of social interaction. In: Lamb, M. E., Suomi, S. J., and Stevenson, G. R. (eds.), *Social Interaction Analysis: Methodological Issues*, University of Wisconsin Press, Madison, pp. 291—309.

Milton, K., and May, M. L. (1976). Body weight, diet, and home range area in primates. *Nature (Lond.)* **259**:459—462.

Modahl, K. B., and Eaton, G. G. (1977). Display behavior in a confined troop of Japanese macaques (*Macaca fuscata*). *Anim. Behav.* **25**:525—535.

Morris, K., and Goodall, J. (1977). Competition for meat between chimpanzees of the Gombe National Park. *Folia Primatol.* **28**:109—121.

Oates, J. F. (1977). The guereza and its food. In: Clutton-Brock, T. H. (ed.). *Primate Ecology: Studies of Feeding and Ranging Behavior in Lemurs, Monkeys, and Apes*, Academic, London. pp. 275—321.

Oates, J. F., Swain, T., and Zantovska, J. (1977). Secondary compounds and food selection by Colobus monkeys. *Biochem. Syst. Ecol.* **5**:317—321.

O'Donald, P. (1962). The theory of sexual selection. *Heredity* **19**:541—551.

O'Donald, P. (1963). Notes and comments: sexual selection and territorial behavior. *Heredity* **18**:361—364.

Packer, C. R. (1977). Inter-troop transfer and inbreeding avoidance in *Papio anubis* in Tanzania. Doctoral Thesis, University of Sussex.

Packer, C. R. (1979). Male dominance and reproductive activity in *Papio anubis*. *Anim. Behav.* **27**:37—45.

Partridge, L. (1978). Habitat selection. In: Krebs, J. R., and Davies, N. B. (eds.), *Behavioral Ecology: An Evolutionary Approach*, Sinauer, Sunderland, Mass., pp. 351—376.

Patterson, I. J. (1975). Aggressive interactions in flocks of rooks *Corvus frugilegus* L.: A study in behavior-ecology. In: Baerends, G., Beer, C., and Manning, A. (eds.), *Function and Evolution in Behavior*. Clarendon, Oxford, pp. 169 183.

Petrinovich, L. (1981). A method for the study of development. In: Immelmann, K., Barlow, L. W., Main, M., and Petrinovich, L. (eds.), *Behavioral Development: The Bielefeld Interdisciplinary Project*, Cambridge University Press, Cambridge.

Pittendreigh, C. S. (1958). Adaptation, natural selection, and behavior. In: Roe, A., and Simpson, G. G. (eds.), *Behavior and Evolution*, Yale University Press, New Haven, pp. 390—416.

Pringle, J. W. S. (1951). On the parallel between learning and evolution. *Behavior* **3**:174—215.

Quick, H. F. (1974). *Population Ecology*, Pegasus, New York.

Rasmussen, D. R. (1973). Spatial relations among adult male Japanese macaques. *Primate News* **11**:11—15.

Rasmussen, D. R. (1978). Environmental and behavioral correlates of changes in range use in a troop of yellow (*Papio cynocephalus*) and a troop of olive (*P. anubis*) baboons. Unpublished Doctoral Dissertation, University of California, Riverside.

Rasmussen, D. R. (1979). Correlates of patterns of range use of a troop of yellow baboons (*Papio cynocephalus*): I. Sleeping sites, impregnable females, births, and male emigrations and immigrations. *Anim. Behav.* **27**:1098—1112.

Rasmussen, D. R. (1980a). Communities of baboon troops (*Papio cynocephalus*) in Mikumi National Park, Tanzania: a preliminary report. *Folia Primatol.* (in press).

Rasmussen, D. R. (1980b). Clumping and consistency of primates' patterns of range use: definitions, sampling, assessment, and applications. *Folia Primatol.* (in press).

Rasmussen, D. R. (1980c). Pair-bond strength and stability and reproductive success. *Psych. Rev.* (in press).

Rasmussen, D. R., and Rasmussen, K. L. (1979). Social ecology of adult males in a confined troop of Japanese macaques (*Macaca fuscata*). *Anim. Behav.* **27**:434–445.

Reinhardt, W., Mutiso, F. M., and Reinhardt, A. (1978). Resting habits of zebu cattle in a nocturnal enclosure. *Appl. Anim. Ethol.* **4**:261–271.

Rodman, P. S. (1973). Synecology of Bornean primates. I. A test for interspecific interactions in spatial distribution of five species. *Am. J. Phys. Anthropol.* **38**:655–660.

Rowell, T. E. (1966). Hierarchy in the organization of a captive baboon group. *Anim. Behav.* **14**:430–443.

Rowell, T. E. (1967a). Variability in the social organization of primates. In: Morris, D. (ed.), *Primate Ethology*. Doubleday & Co., Garden City, pp. 283–305.

Rowell, T. E. (1967b). A quantitative comparison of the behavior of a wild and a caged baboon group. *Anim. Behav.* **15**:499–509.

Rumbaugh, D. M. (1970). Learning skills of anthropoids. In: Rosenblum, L. A. (ed.), *Primate Behavior: Developments in Field and Laboratory Research*, Vol. 1, Academic, New York, pp. 1–70.

Russell, W. N. (1959). Evolutionary concepts in behavioral science: II. *Yr. Bk. Gen. Syst. Res.* **4**:45–73.

Sade, D. S. (1972). Sociometrics of *Macaca mulatta*. I. Linkages and cliques in grooming matrices. *Folia Primatol.* **18**:191–223.

Schneirla, T. C. (1950). The relationship between observation and experimentation in the field study of behavior. *Ann. N. Y. Acad. Sci.* **51**: Article 6.

Schneirla, T. C. (1965). Aspects of stimulation and organization in approach/withdrawal processes underlying vertebrate behavioral development. *Adv. Stud. Behav.* **1**:1–74.

Seligman, M. E. P., and Hager, J. L. (eds.) (1972). *Biological Boundaries of Learning*, Appleton-Century-Crofts, New York.

Skinner, B. F. (1975). The shaping of phylogenetic behavior. *J. Exp. Anal. Behav.* **24**:117–120.

Skinner, B. F. (1976). *About Behaviorism*, Vantage, New York.

Slobodkin, L. B. (1978). Is history a consequence of evolution? In: Bateson, P. P. G., and Klopfer, P. H. (eds.), *Perspectives in Ethology*, Vol. 3, Plenum, New York, pp. 233–255.

Southwick, C. H., Mirza, A. B., and Siddiqi, M. R. (1965). Rhesus monkeys in North India. In: DeVore, I. (ed.), *Primate Behavior Field Studies of Monkeys and Apes*, Holt, Rinehart, & Winston, New York, pp. 111–159.

Southwick, C. H., Siddiqi, M. F., Farooqui, M. Y., and Pal, B. C. (1976). Effects of artificial feeding on aggressive behavior of rhesus monkeys in India. *Anim. Behav.* **24**:11–15.

Spilerman, S. (1975). Forecasting social events. In: Land, K. C., and Spilerman, S. (eds.), *Social Indicator Models*, Russell Sage Foundation, New York, pp. 381–403.

Struble, R. G., and Riesen, A. H. (1978). Changes in cortical dendrite branching subsequent to partial social isolation in stumptailed monkeys. *Devel. Psychobiol.* **11**:479–486.

Struhsaker, T. T. (1969). Correlates of ecology and social organization among African cercopithecines. *Folia Primatol.* **11**:80–118.

Struhsaker, T. T. (1974). Correlates of ranging behavior in a group of red colobus monkeys (*Colobus badius tephrosceles*). *Am. Zool.* **14**:177–184.

Teleki, G. (1973). *The Predatory Behavior of Wild Chimpanzees*, Bucknell University Press, Lewisburg, Pa.

Tinbergen, N. (1951). *The Study of Instinct*, Clarendon, London.

Tinbergen, N. (1963). On aims and methods of ethology. *Z. Tierpsychol.* **20**:410–433.

Tinbergen, N. (1965). Behavior and natural selection. In: Moore, J. A. (ed.), *Ideas in Modern Biology*, Natural History, New York, pp. 521–542.

Tinbergen, N. (1968). On war and peace in animals and man. *Science* **160**:1411–1418.

Tinbergen, N. (1969). Ethology. In: Harre, R. (ed.), *Scientific Thought 1900-1960*, Clarendon, Oxford, pp. 238–268.

Tinbergen, N. (1972). Functional ethology and the human sciences. *Proc. R. Soc. London B* **182**:385–410.

Trivers, R. L. (1971). The evolution of reciprocal altruism. *Q. Rev. Biol.* **46**:35–57.

Trivers, R. L. (1972). Parental investment and sexual selection. In: Campbell, B. (ed.), *Sexual Selection and the Descent of Man*, Aldine, Chicago, pp. 136–179.

Trivers, R. L. (1974). Parent-offspring conflict. *Am. Zool.* **14**:249–264.

Waddington, C. H. (1957). *The Strategy of Genes*, Allen & Unwin, London.

Washburn, S. L., and DeVore, I. (1961). Social behavior of baboons and early man. *Vik. Fund Publ. Anthropol.* **31**:91–104.

Wecker, S. C. (1963). The role of early experience in habitat selection by the prairie deer mouse, *Peromyscus maniculatus bairdi. Ecol. Monogr.* **33**:307–325.

West-Eberhard, M. J. (1975). The evolution of social behavior by kin selection. *Q. Rev. Biol.* **50**:1–33.

Williams, G. C. (1966). *Adaptation and Natural Selection: A Critique of Some Current Evolutionary Thought*, Princeton University Press, Princeton, N.J.

Wilson, E. O. (1975). *Sociobiology: A New Synthesis*, Harvard University Press, Cambridge, Mass.

Wilson, S. F. (1979). Habitat design and behavioral management of apes in captivity. Paper presented to the Western Psychological Association.

Wright, S. (1945). Tempo and mode in evolution: a critical review. *Ecology* **26**:415–419.

Wrangham, R. W. (1977). Feeding behavior of chimpanzees in Gombe National Park, Tanzania. In: Clutton-Brock, T. H. (ed.), *Primate Ecology: Studies of Feeding and Ranging Behavior of Lemurs, Monkeys and Apes*, Academic, London, pp. 504–538.

Wynne-Edwards, V. C. (1962). *Animal Dispersion in Relation to Social Behavior*, Oliver & Boyd, Edinburgh.

Zahavi, A. (1975). Mate selection — a selection for a handicap. *J. Theor. Biol.* **53**:205–214.

Chapter 5

SOCIAL STRUCTURE AND INDIVIDUAL ONTOGENIES: PROBLEMS OF DESCRIPTION, MECHANISM, AND EVOLUTION

R. Haven Wiley

Department of Zoology
University of North Carolina
Chapel Hill, North Carolina 27514

I. ABSTRACT

An old problem in the study of societies is the relationship between social structure and the development of individuals. Societies generally have a more or less stable distribution of individuals among social positions in spite of the passage of successive generations of individuals. For biologists, the problem of the relationships between individual and society has three aspects: the quantitative description of individuals' movements through successive social positions; the mechanisms that regulate this movement; and the evolution of these regulatory mechanisms.

The concept of "ontogenetic trajectories" can serve as a basis for describing the movements of individuals through social positions. Actual measurements of the age-dependent rates of transitions of individuals from one social position to another should soon permit the first relatively complete descriptions of complex animal societies. Ontogenetic trajectories raise questions about the mechanisms that generate stable distributions of individuals among social positions. In some cases, there is evidence of feedback control in the form of inhibition by individuals in a later social position on the development of individuals in earlier positions. In other cases, stable social structure might result from relatively constant demographic conditions and developmental schedules.

The evolution of ontogenetic trajectories, in particular ones that involve substantial deferment of successful reproduction, is related to the

evolution of life-history strategies. Several recurring forms of social structure in birds and mammals, in particular polygyny and stable groups of
cooperating adults with one reproductive pair, raise similar questions about
the evolution of maturational controls that result in ontogenetic trajectories
with delayed reproduction. Some simple calculations can show that the
evolution of delayed reproduction depends on the consequences of early
reproduction for subsequent survival and fecundity. In the end, explanations for the evolution of polygyny and cooperative groups will need to incorporate ecological explanations for the adaptedness of ontogenies with
delayed reproduction, rather than present delayed reproduction as a secondary consequence of the social structure.

II. INTRODUCTION

In the first half of this century, those biologists who turned their atten-
tions to animal societies often adopted simple analogies for the relationship
of society and individual. Students of social insects likened the societies of
ants, bees, and wasps to individual organisms. Indeed, the analogy between
society and organism, an idea with a long history, was particularly cap-
tivating for biologists during the second quarter of the century (see Comte,
1830; Mills, 1843; Wheeler, 1926; Canon, 1932; Clements, 1935; Emerson,
1939; Jennings, 1942). There undoubtedly exist certain fundamental parallels in the relationships between any two levels of biological organization.
Understanding how cells interact to regulate an organism's internal
homeostasis has some similarities, at least in form if not content, to understanding how individual animals interact to regulate a population's social
structure. Nevertheless, there are some crucial differences, which few
biologists fully perceived until recently.

Already in 1932, Haldane had recognized a key problem in the evolution of individuals' relationships in a society. How could natural selection
favor the genes of individuals, such as the sterile workers in the social insects, that sacrificed their own reproduction and increased the reproduction
of others? This question of the evolution of altruism, neglected for three
decades, was to provide a major impetus for the recent development of a
biology of animal societies.

Intensive biological investigation of the relationships of individuals and
society began around 1930 with Allee's (1942) studies of dominance
hierarchies and Schneirla's (1971) studies of army ants. Both focused on the
behavioral and physiological mechanisms that regulate social structure. Allee and his students investigated in detail the physiological and experiential

determinants of dominance, but they also took a clear interest in the effects of dominance on the reproduction and survival of individuals (Allee, 1942). They concluded that the greater fecundity and physiological condition of dominant animals indicated greater evolutionary fitness. Thus, prior to 1950, biologists had taken the first steps in understanding the behavioral and physiological control of social structure and had identified some central problems in the evolution of social behavior.

During this period, sociologists who wrestled with the relationships between individuals and society had little discernible influence on biology. Some schools of sociology emphasized the effects of society on the behavior and development of individuals. Durkheim (1972) in particular elaborated the view that every society has a collective conscience that persists across successive generations of individuals and molds the behavior of each generation. Social psychologists, following Mead (1934) and Merton (1949), emphasized that individuals appraise their actions in the light of the expectations of others. Other schools of sociology gave more emphasis to the effects of individual motivation on social behavior. Parsons (1951), for instance, attempted to reconcile psychological and sociological approaches to human behavior by developing a conceptual scheme in which human action is controlled by interacting social, cultural and motivational influences.

Perhaps one reason that sociology and biology have found so little common ground stems from a procedural difficulty. Sociologists share, in spite of their differences, a common emphasis on a system of norms or expectations held in common by sets of individuals in a society. The concept of norms of behavior is not easily accommodated in studies of animal societies. Sociologists determine the norms or expectations of the members of a society by means of verbal interaction with their subjects, or in some cases introspection — techniques that have no application in the study of animal societies. Even the recent advances in verbal communication with apes raise no immediate hope of verbal interactions with apes in their naturally occurring societies.

Nevertheless, the basic problem, if not the technique, that has occupied sociologists seems applicable to animal societies. A characteristic of all societies is a recognizable structure that persists with relative constancy in spite of the passage of succeeding generations of individuals. At any one time, the individuals in a particular population exhibit characteristic differences in behavior, often identifiable as more or less discrete behavioral categories. The basic characteristics of the behavior of individuals in these categories and the proportions of the population or the absolute numbers of individuals in each category often remain relatively constant with time. This structure of a society, in spite of some important variation in time and space, remains constant enough that we can associate characteristics of social structure with particular populations or species of animals. The persistence of social structure in spite of a continual turnover of individuals has prompted such concepts in sociology as a collective conscience or norms of behavior. Although these concepts are not useful in analyzing animal societies, the question remains a legitimate one: how do successive generations of individuals come to adopt such similar social organizations?

A biologist immediately sees that this question presents a host of problems in the genetics, physiology, and development of behavior. Although many relevant aspects of these problems have received attention, biologists have yet to put this information together in a coherent explanation of the mechanisms that control the relationships between individual behavior and social structure for any species. During the spate of research since 1960 on the evolution of animal societies, there has been no comparable attempt to develop a systematic framework for understanding the mechanisms of animal societies, the physiological and behavioral controls of individuals' ontogenies that generate a persistent social structure.

Before analyzing the mechanisms that regulate the social structure of a population, we need a framework for describing the flow of individuals in time through positions in the society. By analogy, in physics a study of kinematics, a quantitative description of the movements of the components of a system without regard to the forces involved, might precede a study of dynamics. To describe the flow of individuals in time through a more or less persistent social structure, I have found helpful the concept of "ontogenetic trajectories" of individuals through social positions in a society. A trajectory is an appropriate analogy for the movement of individuals through positions in a society. The development of any individual depends on its initial genetic endowment and its exposure to environmental and social influences during the course of its life, just as the movement of a projectile depends both on its initial conditions and its continuing interaction with the medium through which it passes.

Biologists attempting to understand the relationship between individuals and societies thus confront three categories of problems: description, mechanism, and evolution. The following pages develop the concept of ontogenetic trajectories first as a tool for describing the structure of a society, then as a basis for analyzing the behavioral and physiological mechanisms that regulate social structure, and finally as a perspective on animal societies that raises some pervasive evolutionary questions.

III. ONTOGENETIC TRAJECTORIES

Although the development of behavior has long been a major emphasis in psychology and ethology, most of this work has focused on the development of behavior in immature animals. The concept of ontogenetic trajectories can serve to emphasize the important changes in the social positions of individuals throughout their lives.

Changes in the social behavior of individuals are probably best understood in social insects. The most thoroughly documented example is the worker honeybee. By following marked individuals, an observer finds not only a general change from participation in duties within the hive to foraging outside the hive, but also more subtle changes in the kinds of duties that occupy bees within the hive (Lindauer, 1952; Wilson, 1971).

This example probably provides the only case so far in which ontogenetic trajectories have been determined directly, by observing marked individuals throughout their lives. Indirect determinations of ontogenetic trajectories rely on observations of individuals changing from any one social position to others. Individuals' possible ontogenetic trajectories are then reconstructed from these observations of changes from one social position to the next. This sort of reconstruction amounts to the use of a first-order Markov process to estimate sequences of behavior. In this case each state is a social position, and the observer assumes that transitions of individuals from one social position to another occur independently of their previous histories. This indirect analysis of ontogenetic trajectories has been applied extensively to the castes in termite societies.

The movement of individuals through different castes in termite societies illustrates some of the features of ontogenetic trajectories (see Wilson, 1971; McMahan, 1979). In the primitive families of termites, the changes of individuals from one caste to another are extremely complex. Often colonies include individuals with characteristics intermediate between the definitive characters of different castes. Furthermore, individuals have great developmental flexibility. At any age after the third or fourth instar, individuals can molt into soldiers or reproductives and even molt retrogressively (Miller, 1969).

In contrast, most of the higher termites (Termitidae) have relatively well-defined ontogenetic trajectories of individuals through different castes (Noirot, 1969). Often the trajectories permit no branching or regression after the first instar. Particularly interesting, however, are trajectories that branch into two discrete paths (Fig. 1).

Although the paths by which individuals develop are now well established for a variety of social insects, much less is known about the time that individuals spend in successive social positions or castes. Estimates of these durations and their variances require observations of marked individuals. When individuals cannot be followed throughout their lifetimes, an observer can follow samples of marked individuals in each identifiable social position. A map of the possible ontogenetic trajectories of individuals through social positions in combination with estimates of the rates at which

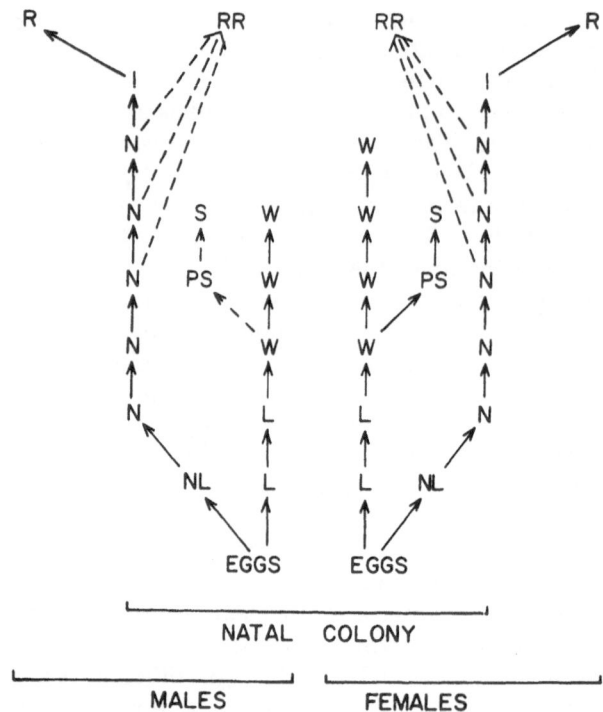

Fig. 1. Ontogenetic trajectories of a higher termite (*Microcerotermes*, Termitidae). The vertical scale indicates successive instars, not actual time; the stages vary in duration.
Developmental paths for reproductive and nonreproductive individuals divide at an early age,
possibly before the eggs hatch. Larvae are indistinguishable by external morphology in the first
stage, but before the first molt they have already divided into two groups: larvae (NL) destined
to develop into nymphs (N), imagoes (I), and reproductives that found new colonies (R) or
replacement reproductives in the natal colony (RR); and larvae (L) destined to develop into
workers (W) and soldiers (S). In this genus, female nonreproductives are larger than males, a
difference that becomes evident among second-stage larvae. Such differences in size often correlate with behavioral differences. Soldiers develop primarily from female workers, rarely from
males. In other genera of higher termites, ontogenetic trajectories of nonreproductives take a
variety of patterns, but males and females usually have substantial differences. In the families
of lower termites, developmental pathways are not so fixed. In particular, the early division
between reproductive and nonreproductive lines does not occur and there are no differences
between the sexes in ontogenetic trajectories. Modified after Noirot (1969).

individuals change from one position to the next (or the residence times of
individuals in each social position), the mortality rates of individuals in each
social position, and the recruitment of individuals to the population provide
the information for a complete kinematic model of a society.

Among vertebrates most species have societies with much less complex structure than those of termites. In many cases, societies seem to have only a few identifiable social positions for mature animals. Consider a territorial passerine bird. During the breeding season, we might recognize territorial and nonterritorial males and mated and unmated females: two social positions for all individuals of each sex.

Of course the structural complexity of a society depends on how finely we classify social positions. We could recognize several categories of behavior for a territorial male bird, for instance feeding, courting females, and defending his territory; then we could analyze the individual's transitions from one to another of these categories of behavior on a finer timescale than we could analyze longer-term changes from nonterritorial to territorial status. In analyzing social structure of vertebrates, I will use the term social position to refer to patterns in individuals' social behavior that normally persist over periods of days at least, in order to distinguish these patterns from those that are recognizable over periods of minutes or hours.

The term social "role" has in the past applied to both of these time scales. "Role" has an additional ambiguity as a result of a somewhat different connotation in much sociological literature. There it often means the norms or expectations for an individual's actions, while "position" usually refers to more objective criteria for differentiating individuals, such as an individual's age, title, genealogical relationships, or actions. Sociologists do not always draw this distinction, but when they do, the term "position" is more consistently used for the objective criteria that differentiate individuals (Sarbin and Allen, 1969).

The different species of grouse (Tetraonidae) provide some instructive contrasts in complexity of social structure. According to the thorough studies of red grouse (*Lagopus lagopus*), males older than about six months fall into only two categories of social behavior, territorial and nonterritorial birds. Males barely four months old seem equal to older males in obtaining territories and mates (Watson and Moss, 1972). Thus the red grouse fits the pattern for a simple avian society with two social positions for mature individuals of each sex. In contrast, among those species in which the males congregate at leks, all full-grown males are not equivalent in their access to choice territories. Those males with territories near the center of a lek copulate much more frequently with visiting females than do males with peripheral territories or nonterritorial males (reviewed by Wiley, 1973).

Just in the last decade evidence has accumulated that males usually acquire central territories after first establishing peripheral territories on a lek and then moving their positions toward the center. I inferred this centripetal movement of males on leks of sage grouse (*Centrocerus urophasianus*) by the indirect means of observing the reoccupation of territorial vacancies (Wiley,

1973). Invariably a vacancy was occupied by one or more neighboring males who moved their territories toward the center of the lek. This sort of indirect evidence is much like that used to deduce the trajectories of termites through castes. More direct evidence comes from studies of marked populations. Both for sharp-tailed grouse and black grouse, males on the periphery of a lek have moved subsequently into central positions (Kruijt et al., 1972; de Vos, 1979; Rippin and Boag, 1974). No study yet has marked enough birds and followed them long enough to establish the means and variation for the transition rates between positions on a lek or the residence times in different positions on a lek.

It seems possible that the males of many polygynous species have ontogenetic trajectories that involve a sequence of social positions leading to a position that permits high fecundity, in essence a queue for access to opportunities for reproduction. The lek-forming grouse provide an example in which such ontogenetic trajectories correspond with actual movements of individuals' territories in space. Probably in all polygynous birds and mammals, males begin successful reproduction on average at later ages than do females. Among antelope, males acquire territories and opportunities to copulate with females several years after females begin reproduction (see, e.g., Spinage, 1969; Jarman and Jarman, 1974; Gosling, 1974). Male elephant seals (*Mirounga angustirostris*) continue to grow for approximately twice as many years as do females. Although only the larger, and consequently older, males become harem masters, males begin to frequent breeding grounds regularly at younger ages (LeBoeuf, 1974). One wonders whether queuing might not also occur in these species, in the sense that a male acquires a position as a successful breeder through a prolonged and gradual process of exploiting small advantages among males in the immediate area in which he eventually succeeds to full reproductive status.

Among some primates also, older males tend to dominate younger ones, and more dominant males tend to copulate most frequently, although there is considerable variation among species and even among populations in both of these generalizations (Hausfater, 1975; Wade, 1978; Packer, 1979). Primates, owing to their relatively long lives, have been difficult subjects for descriptions for social kinematics. Until very recently, most studies of primate societies in the field reported only isolated anecdotes of mature individuals that changed social positions. The histories of dominance and reproductive success of males throughout their lives have been especially difficult to obtain. The initiation in the last two decades of long-term studies of primates in the field should remedy this situation soon.

Another complex form of vertebrate society, stable groups of cooperating adults, also invites an analysis of ontogenetic trajectories. Several mammalian carnivores, in particular the wolf (*Canis lupus*), the hunting dog (*Lycaon pictus*), and a few viverrids, share some striking features of social behavior with certain birds in more than a dozen phylogenetic families (Brown, 1978; Emlen, 1978; Rasa, 1977; Frame and Frame, 1976; Mech, 1970). In these species, adults associate in stable groups that usually average five to eight. The members of a group cooperate in defending a shared territory. Usually only one male and female in each group breeds, although the other members of the group help to feed the young. These groups thus consist of two clearly different social positions for adults, reproductives and nonreproductives. For the study of ontogenetic trajectories, the birds have some clear advantages over the mammals with this sort of social organization. Because large numbers of birds can often be marked individually and then be subject to a reliable census, information accumulates more rapidly on the transition rates and residence times of individuals in different social positions.

The South American stripe-backed wren (*Campylorhynchus nuchalis*) presents optimal conditions for a complete documentation of the ontogenetic trajectories in a complex vertebrate society. The members of a group of stripe-backed wrens all sleep each night in the same enclosed nest, a habit that makes it possible to take a complete census of the population by counting the birds as they leave and enter their nests. In addition, by decoying the birds with playbacks of tape-recorded vocalizations, one can capture and mark individually almost every individual in a population. Rabenold now has more than 150 individuals in 25 contiguous territories under study (see Wiley and Wiley, 1977; Wiley, 1978; Rabenold and Christensen, 1979).

To construct a kinematic model of this society, we recognize three social positions: dependent juveniles, auxiliary (nonreproductive) members, and principal (reproductive) members of groups. The preliminary data are sufficient to give a general picture of the model; Rabenold will soon have estimates of the transition rates between social positions and residence times for each sex in each social position. The picture that emerges is a remarkably general pattern of social kinematics for species with stable groups in which one male and female reproduce and other adults help (Fig. 2). A sex difference in ontogenetic trajectories applies to hunting dogs as well as the group-territorial birds: females become reproductives primarily through emigration from their natal groups, whereas males become reproductives within their natal groups (Zahavi, 1974; Frame and Frame, 1976; Woolfenden and Fitzpatrick, 1978). For males, the process of ascending to reproductive status amounts to a form of queuing. Even for females, at least in stripe-backed wrens, there is an element of queuing in succession to reproductive status, since a vacancy for a female breeder is usually filled by a female from an immediately adjacent territory. In both sexes, it pays to have lived in the immediate area when an individual tries to succeed to a reproductive position.

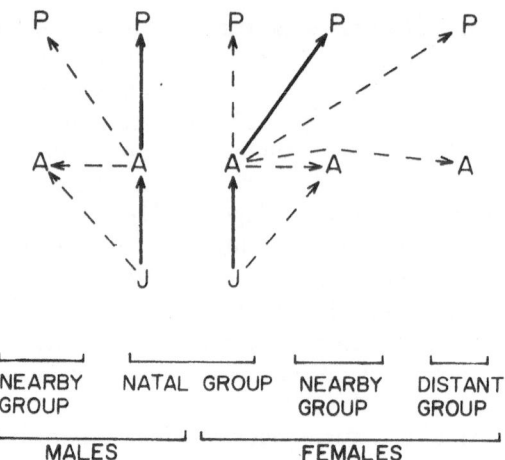

Fig. 2. Ontogenetic trajectories of stripe-backed wrens (*Campylorhynchus nuchalis*, Troglodytidae). The vertical scale indicates successive social positions in stable social groups; the stages vary in duration. Most juveniles (J) join their natal groups as nonreproductive adults or auxiliaries (A). A female usually succeeds to a reproductive, or principal (P), position by occupying a vacancy created by the disappearance of a principal female in a nearby group. A male usually becomes a principal member of a group by succession within the natal group. Dashed lines indicate infrequent transitions between social positions. Based on studies by K. Rabenold, C. Christensen, M. and R. H. Wiley.

IV. HOMEOSTASIS OR STEADY STATE?

Quantitative description of the ontogenetic trajectories of individuals in a society immediately raises questions about the behavioral and physiological mechanisms that regulate these trajectories. To pursue our analogy with physics, we need to move from the kinematics to the dynamics of the movements of individuals through social positions.

At the outset of this review, we noted that one of the remarkable features of the relationship between individuals and societies is a relative constancy in the proportions or numbers of individuals in each social position in spite of the continual turnover of individuals. Such stable structure could arise either with or without feedback control of the recruitment of individuals to successive social positions. Feedback would consist of an inhibition of the recruitment of individuals to a particular social position as a function of the number of individuals already occupying that position. For instance, the presence of reproductive animals could inhibit the development of reproductive behavior in nonreproductive animals.

A stable social structure, however, need not require regulation by such feedback. If reproduction and mortality, along with the characteristics of individuals' ontogenetic trajectories, remained relatively constant over periods of many years, the social structure of such a population would reach a steady state. Any brief perturbations of the demography of such a population would result in corresponding alterations in social structure. The population would regain a steady social structure only over periods approximating the lifetimes of individuals and the durations of environmental perturbations. In contrast, feedback regulation of the succession of individuals to social positions would assure some homeostasis of the social structure in the face of demographic perturbations of the population.

In many social insects, the number of reproductives in a colony is under strong feedback control, mediated either by direct aggression of current reproductives against nonreproductives or by pheromones. Accumulating evidence also suggests that the ratio of soldiers to workers is under homeostatic control in some species (Wilson, 1971; Brian, 1979). No vertebrate society has received a comprehensive analysis of the mechanisms that regulate individuals' ontogenetic trajectories through social positions. Research on reproductive maturation in mammals, however, clearly establishes the possibility of feedback regulation of succession to reproductive social positions. In both male and female house mice, exposure to older reproductive individuals of the same sex inhibits reproductive maturation (Vandenbergh, 1974; Bronson, 1974). Exposure to reproductive individuals of the opposite sex, on the other hand, accelerates maturation. In house mice, both of these effects are stronger in females than in males. In fact, the inhibition of reproductive maturation by exposure to older males has a relatively slight, although reliable, effect on the maturation of younger males. In house mice these effects can result entirely from persistent olfactory stimuli in urine from reproductive animals.

Even if interactions with reproductive individuals fail to inhibit the maturation of younger individuals, resident animals might still prevent the establishment of young recruits in the population. This possibility has received a great deal of attention in discussions of territoriality in birds. At least in some species, established territorial residents exclude other individuals from optimal habitats. In some cases, like the great tit (*Parus major*) and the chaffinch (*Fringilla coelebs*), the excluded individuals find territories in less suitable habitats (Glass, 1960; Krebs, 1971). In red grouse, many excluded individuals fail to obtain territories at all. In this species, as we have noted, success or failure in obtaining a territory seems to have no relation to an individual's age. There is no ontogenetic progression from nonterritorial to territorial status or from suboptimal to optimal habitats. In other species though, young males are more common in suboptimal

habitats. In great tits, Krebs (1971) observed that males originally established in suboptimal habitats, hedgerows, moved their territories into optimal habitats, continuous woodland, when vacancies became available
there.

Anecdotal information in the literature does not always allow a clear
decision about the possibilities of feedback from social interactions in
regulating the succession of individuals to reproductive status. The situation
is particularly unclear for many polygynous species. In polygynous species,
males begin to reproduce on average at a later age than do females. In many
cases, they also reach full morphological and physiological development at a
later age. For instance in polygynous birds like great-tailed grackles
(*Quiscalus mexicanus*), red-winged blackbirds (*Agelaius phoeniceus*), and
lek-forming species of grouse, males one year old have less developed
plumage than older males, although they have already reached full weight.
In addition the gonads of one-year-olds mature later in the breeding season
than do older males' and never reach the full size of older males' (Wright
and Wright, 1944; Selander and Hauser, 1965; Eng, 1963; Payne, 1969;
Selander, 1972). In these species, one-year-old males rarely obtain territories
suitable for reproduction.

On the basis of experiments in which older males have been removed
from their territories and some young males have occupied their places, it is
tempting to conclude that social interactions with older males inhibit
reproductive development and behavior of young males. In fact, more careful examination of the evidence from removal experiments and some recent
experiments with captive blackbirds suggest that, at least for grouse and
blackbirds, social inhibition is unlikely to explain why young males fail to
reproduce. In these blackbirds and in the lek-forming species of grouse,
young males arrive on breeding grounds later in the season than do older
males, in accordance with the later seasonal development of their gonads
(Orians, 1961; Payne, 1969). In fact, for sage grouse, first-year males initially
begin territorial behavior at the periphery of a lek about the time that
copulations begin at the center, at least a month after the older males have
started territorial activities. The territorial behavior of many first-year males
barely overlaps with the period of copulation (Wiley, 1973).

In an experiment reported in the Russian literature (see Dementiev and
Gladkov, 1967), Kirikov tried removing older males from leks of capercail-
lie (*Tetrao urogallus*) to see if younger males would establish territories
earlier in the season. He reports that the presence of older males had no
effect on the arrival dates of younger males, although in the absence of older
males the younger males occupied the center of the lek. Thus although
younger males occupied vacancies left on leks by older males, they only did
so later in the season than older males normally began territorial activity.

Similar findings are reported for red-winged blackbirds. Again, although younger males will occupy vacancies left by the removal of older territorial males, they only do so late in the season after mating has begun for the females' first clutches (Orians, 1961). Earlier in the season, vacancies are reoccupied by other old males.

Thus the delay in maturation of the gonads and territorial behavior of first-year males is the critical factor in the regulation of their access to reproductive positions. In order to determine whether or not the seasonal course of physiological and behavioral maturation of first-year male red-winged blackbirds depends on direct social interactions with older males, Hartnett and I studied small groups of males in outdoor aviaries (Wiley and Hartnett, 1976). We compared three kinds of groups: three first-year males, one older male with two first-year males, and three older males. In each group one of the males became clearly dominant over the other two; in groups with mixed ages, the older male invariably became dominant. Thus we could compare the physiological and behavioral development of a dominant first-year male and dominant older males, all exposed to two subordinate first-year males. The dominant first-year males became as aggressive toward their subordinates as the older dominants did, but the increase in aggressive behavior proceeded more slowly during early spring than for the older males. As in the uncontrolled situation in the field, the dominant younger males became fully aggressive only in early May, about the time that most of the females lay their first clutches. Furthermore, the dominant young males had significantly smaller testes than the dominant older males in similar social situations. Dominance rank and exposure to older males thus had no major effects on the maturation of the gonads of first-year males. Contrary to our initial expectations, the behavioral and physiological development of young red-winged blackbirds seems largely independent of social interactions with older males.

So far we have focused on the mechanisms that regulate or generate the typical structure of a society. For this purpose we have concentrated on mechanisms that control the average parameters of the ontogenetic trajectories of individuals in the society. It is equally important to identify the factors that control the variation in ontogenetic trajectories. Although this variation would have little effect on the long-term or equilibrium structure of the society, it will have important effects in determining which of the available individuals eventually succeed to positions that permit successful reproduction.

For example, although most first-year male sage grouse establish territories around the peripheries of leks, there is substantial variation in the time during the season when they first become effectively territorial. Furthermore, among those males that survive to their second year, there is undoubtedly substantial variation in the progress they make toward the center of the lek and successful reproduction. This sort of variation in ontogenetic trajectories depends on the relationships of individuals that occupy the same social position rather than, as before, the relationships of individuals occupying different social positions.

Fights between individuals provide one mechanism by which some individuals advance while others do not. Although direct aggression has undoubted importance as the court of last resort in deciding such matters, fighting is often nearly absent during the crucial movements of individuals into more advantageous positions. Sage grouse fight on their leks, occasionally severely, but these fights occur between established territorial males when one has transgressed its boundary. In contrast, when a vacancy occurs on a lek there is remarkably little fighting. In fact, the territory of a male that has suddenly disappeared often remains unoccupied for one or two days before a peripheral neighbor gradually begins to extend his activities into the vacancy. Even when two peripheral neighbors, to all appearances, have equal claims to the vacancy, there is no noticeable aggression. Similarly, in early spring when the first males begin to occupy positions on the lek, more than a month before the females arrive, agonistic interactions are infrequent and lackadaisical. At this time of year, toward the end of winter, leks are likely to have more vacancies than at any other time, and thus males should have the greatest opportunities for improving their positions.

Such observations suggest that differences in the advancement of individual males can depend more on their endurance and their luck than on direct aggression. Physiological endurance could have major effects here. Attendance at a lek in early spring requires some sacrifice in opportunities to feed. Those that can afford the necessary time at a lek might well stand to advance their positions. However, each male must balance these advantages against the possible disadvantages that would accrue from poor physiological condition as a result of insufficient time for feeding. Similar considerations might apply both to older males early in the spring, when snow is first disappearing, and to first-year males when they establish their territories for the first time toward the middle of the season. In general, similar considerations must apply to all individuals at the time they commence territorial behavior (Searcy, 1979b).

We cannot neglect the role of pure chance in determining variation in ontogenetic trajectories. For example, the leks of grouse are rarely exactly symmetrical around the central territories where most copulations occur. Thus the possibility arises that first-year males establishing territories at the periphery of a lek might find themselves at substantially different distances from the center. It would pay for a young grouse to make some assessment of an optimal position to start its career on a lek, but in the absence of such an assessment the young male would have to take the luck of the draw. In addition, the opportunities for advancement presumably depend on the death of territory holders that intervene between a young male's initial position and the center of the lek. Thus the occurrence of vacancies for a young male to occupy is itself a random process that would introduce random variation in the ontogenetic trajectories of males.

Variations in the physiological condition of individuals and the chance availability of opportunities for advancement would generate considerable variation in the ontogenetic trajectories of individuals in any complex society. The importance of fighting in the advancement of individuals through successive social positions, on the other hand, might depend on whether or not orderly queuing can occur for reproductive positions.

Stripe-backed wrens are instructive in this regard, because fierce fighting is usual in advancement to reproductive status by females, but not males. When the reproductive male of a group disappears, one of the older males assumes the vacancy, usually with no signs of conflict within the group. On the other hand, when a reproductive female disappears, her vacancy becomes the subject of intense competition among females from nearby territories. Although invading females often come from as far as four territories away usually a female from a neighboring territory successfully claims the position. The first few days of the contest bring repeated, intense fighting between the presumptive females. Females peck each other viciously and grapple in midair. The auxiliary (nonreproductive) males in a group, like the males on a lek of grouse, participate in a queue for the principal (reproductive) position. Like the sage grouse, succession to reproductive status involves primarily an orderly progression of individuals into the available vacancies. The female stripe-backed wrens that compete for vacant reproductive positions in groups, in contrast, have not queued for this position. Although females from nearby territories have an advantage over those from farther away, the available females from different territories at roughly equivalent distances from a vacancy have had little opportunity for previous encounters. In the absence of opportunities for a slow contest of endurance, the females take the court of last resort and fight it out.

V. EVOLUTION OF MATURATIONAL CONTROLS

Our review of the mechanisms that could regulate individuals' ontogenetic trajectories leads to a question that is fundamental for understanding the evolution of social structure. We can no longer indiscriminately invoke "the good of the species or the group" in explaining the evolution of social behavior, the usual practice in previous generations. The first principle of any acceptable explanation for evolutionary adaptation is the accumulation in a population of those alleles, or combinations of alleles, that spread fastest and persist longest in a particular environment. How then can selection favor the evolution of maturational mechanisms that result in postponement of reproduction, the most direct effect of which is an increase in generation time?

Clearly the evolution of complex social structure in animals is inextricably associated with the evolution of an optimal allocation of effort to reproduction, growth, and maintenance throughout the life of an individual. In spite of the recent wave of interest among evolutionary biologists in the evolution of life-history strategies (Gadgil and Bossert, 1970; Stearns, 1976; Horn, 1978), very little of this approach has penetrated the study of animal societies. Even societies with contrasting structures, like those of stripe-backed wrens and sage grouse, present similar problems in the evolution of ontogenetic trajectories that represent optimal life-history strategies.

The reasons that students of animal societies have so far not fully acknowledged the literature on life-history strategies stem from the wide acceptance of at least three arguments inherited from long before modern advances in evolutionary theory. First is a tacit equation of behavioral competition (aggressive interactions) with evolutionary competition (relative rates of propagation of alleles in a population). Second is an assumption that older animals have an inevitable ability to exclude younger animals from opportunities to reproduce. Third are arguments that fixed behavior by one category of individuals, often one sex, unilaterally determines the evolution of another category of individuals.

First consider the problem of equating behavioral and evolutionary competition. Allee took this step when he inferred that dominant individuals had higher evolutionary fitness than subordinates. However, when we consider cases in which interacting individuals differ in age, the possibility arises that both act in accordance with optimal age-specific strategies, in spite of the fact that one dominates the other. In cases in which evolution favors increasing reproductive effort with age, older individuals should exert more effort in obtaining reproductive social positions than younger individuals. Nevertheless, at any age individuals should always avail themselves of opportunities to reproduce within the constraints of the optimal age-specific effort and risk. One can imagine situations in which individual adaptability in life-history strategy would prove evolutionarily adapted when environmental and demographic conditions changed markedly over periods of approximately an individual lifetime. At any rate, an uneven outcome of behavioral competition between individuals that differ in age could well have no effect on the individuals' relative fitnesses. Uneven behavioral competition between individuals of the same age, on the other hand, would always influence their relative fitnesses.

The assumption that older animals have an inevitable advantage over younger animals in claiming opportunities for reproduction is well entrenched in the literature on the evolution of polygyny. Young males, it is argued, postpone full reproductive development and effort because they cannot compete with older males for mates and consequently have little chance of reproduction. It is true that in polygynous species younger males do not compete successfully with older males for mates. There are two possible explanations for this phenomenon. The traditional one goes as follows: young and old males inevitably differ in competitive abilities; consequently, when males have limited parental duties older males inevitably obtain most of the mates, the population becomes polygynous, and the males evolve delayed maturation. The other explanation runs as follows (Wiley, 1974a): when the fitness of males but not females is increased by delayed reproduction, males evolve deferred reproductive maturation and effort, provided that any consequences of reduced male parental care for breeding success do not counterbalance the advantages of sexual divergence in life histories; consequently, more females than males breed in a given season, which amounts to polygyny. Both explanations depend on ecological conditions that reduce the advantages of full parental care by males (Verner, 1964; Orians, 1969). The first explanation requires an assumption that males cannot evolve full competitive abilities at an early age under any circumstances. The argument hinges on an inevitability of increased competitiveness with age. It thus assumes the consequent. If this assumption of inevitability were not accepted, one would have to explain the evolution of increased competitiveness with age or, the other way around, decreased competitiveness at early ages. In effect, the inevitability assumption is the primary distinction between the two explanations for the evolution of polygyny. In addition to this theoretical objection to the inevitability assumption, the red grouse provides a counter example. In this species, first-year males and older males compete equally for territories. The second explanation above emphasizes two factors in the evolution of polygyny: the adaptedness of life-history strategies of males and the adaptedness of parental strategies.

The two-factor theory of the evolution of polygyny has some similarities to Maynard Smith's (1977, 1978) analysis of parental care. In explaining the evolution of heterosexual relationships, both approaches consider the consequences of the male's strategy as well as the pattern of parental care. Maynard Smith's analysis in terms of game theory focuses on the relative survival of eggs under the care of two, one, or no parents; the tradeoff for a female between producing eggs and guarding them after laying; and the possibilities of additional matings for males that desert their first partners. Maynard Smith considers two strategies for each sex, either caring for the young or deserting. He then establishes the conditions that make each of the four possible patterns of parental care advantageous for each sex, an analysis that elegantly clarifies the interacting influences on the evolution of parental care and mating systems.

This approach converges closely with the two-factor theory for the evolution of polygyny when we consider that the possibilities for additional matings by deserting males are associated with sexual bimaturism, a delay in the age of reproduction by males relative to females. As indicated above, increased fecundity of males in polygynous species is, in large part, associated with delayed reproduction. Thus the analysis must consider the effects of age-specific fecundity and survival on the fitnesses of a male's genes. In the end, a full analysis of the evolution of mating systems will require assessment of the balance between the adaptedness of sexual bimaturism and the adaptedness of different patterns of parental care.

The third problem that impedes applications of life-history theory to social behavior has, like the second, appeared particularly in explanations for the evolution of polygyny. It has often been proposed that females prefer to mate with older males or with males that have already attracted other mates, because such males have demonstrated exceptional abilities to survive or reproduce. The fundamental approach here is sound: females should evolve behavior that results in mating with males that have the highest fitness. However, modern theory of life-history strategies teaches clearly that fitness is not a simple function of survival to old age nor of fecundity in any one season. Females should indeed evolve behavior that results in choosing mates with the highest fitness, but a male's fitness depends on his age-specific allocations of effort to survival and reproduction throughout his life. In this view, the evolution of female preferences converges with the evolution of optimal male life-history strategies. An explanation for the adaptedness of delayed reproduction and high fecundity at a late age among males is basic to understanding the evolution of female preferences in polygynous species.

A deferment of reproductive effort is often not absolute. In polygynous species, for instance, males often initiate a low level of reproductive effort a year or more before they have any real opportunity for success. In fact, these young males sometimes channel their reproductive effort in ways that offer little chance of success even if an opportunity should arise unexpectedly. First-year male sage grouse and red-winged blackbirds, for instance, only mature to full agonistic behavior after the main period of mating has begun. Why do they bother to take any risks of territorial behavior their first year at all?

The answer here could well relate to the queuing effect. It might pay for a first-year male to establish his territorial position, particularly late in the season after the most intense territorial competition has passed, in anticipation of improving his chances for a good territorial position and mating success the next year. On grouse leks, territorial vacancies are usually occupied by established neighbors and almost never by newcomers that arrive abruptly from outside. A male sage grouse probably gets to the head of the line by alertly occupying vacancies as they arise ahead of him. As for red-winged blackbirds, recent evidence suggests that a male's success in attracting females relates in part to having occupied its territory in the preceding year (Searcy, 1979a; Yasukawa, 1979). Exactly why previous occupation of a territory confers an advantage in attracting females is not clear. Possibly males with previously established territories can reoccupy them on average earlier in the season, and perhaps with less agonistic behavior as well, and these two consequences in turn result in greater success in attracting females. At any rate, in these two dissimilar polygynous species young males seem to establish territories late in their first year primarily in anticipation of benefits that accrue for successful mating in subsequent years.

In societies with cooperative breeding, like the stripe-backed wren, most members of a group are usually genealogically related. As a consequence, kin selection influences the evolution of interactions between group members (Brown, 1978; Emlen, 1978). Most of the information available so far about the genealogical relatedness of group members consists of reports that young join their natal group as auxiliary members. Auxiliaries early in life are thus likely to be the progeny of the group's principal members. Few studies, however, have documented the steps by which individuals become principal members of groups. In stripe-backed wrens, a new female principal is normally a newcomer from outside the group and thus usually has low genealogical relatedness to the group's current auxiliaries. A new male principal, on the other hand, is often a sibling or half-sibling of the auxiliaries in the group. Consequently, as an auxiliary member of a group grows older and one or both principals change, its genealogical relatedness to the principals usually decreases. The behavior of an auxiliary, as a result of kin selection, should depend on whether or not the principals have changed since it joined the group at birth. The expected genealogical relatedness of the principals to an auxiliary of any age results directly from the ontogenetic trajectories of individuals in the population.

The evolution of ontogenetic trajectories in societies with cooperative breeding also requires examination in relation to deferred reproduction. The delay in reproduction by the auxiliary members of groups raises the same problems as the delay in reproduction by males in polygynous societies. A full explanation for the evolution of societies with cooperative breeding will thus require consideration of the genealogical relationships of group members, the consequences of participation in groups for the immediate survival and reproduction of individuals in each social position, and the consequences of delayed reproduction (Brown, 1978; Emlen, 1978; Gaston, 1978; Wiley, 1978; Rabenold and Christensen, 1979).

The evolution of deferred reproduction, a critical concern in the preceding discussion, depends on an interaction of several aspects of age-specific survival and reproduction. Evolution of reproduction starting at a given age a will depend on the chances of survival to age a, the expected fecundity at age a, the rate of increase of the population, and the consequences of reproduction at age a for the chances of survival to the next breeding season and for future fecundity.

To see these interacting influences on the evolution of delayed reproductive effort, we can use the standard equation for the rate of population increase,

$$1 = \sum_{a}^{\infty} \lambda^{-x} l(x) m(x)$$

where λ is the proportionate annual population increase (N at $t + 1/N$ at t), $l(x)$ is age-specific survival, $m(x)$ is age-specific fecundity, and a is the age at first reproduction. This equation applies to any lineage of organisms, provided age-specific survival and fecundity remain constant with time and the lineage has had time to reach a stable age-distribution. For diploid organisms, λ provides only an approximate measure of the fitness of an allele associated with a set of age-specific parameters (Charlesworth, 1973). As a first step though, we can ask whether lineages with particular age-specific parameters increase more rapidly than lineages with other parameters. This approach rests on the basic tenet of evolutionary biology: those alleles that propagate fastest accumulate in a population.

Consider two lineages: in one, individuals begin to reproduce at age a; in the other, individuals begin to reproduce at age $a + 1$ and, as a consequence, have higher survival by a factor S between ages a and $a + 1$ and also have higher fecundity by a factor M at age $a + 1$ and thereafter. Thus:

$$l_2(x) = S l_1(x), \qquad x \geqq a + 1, \; S \geqq 1 \tag{1}$$

$$m_2(x) = M m_1(x), \qquad x \geqq a + 1, M \geqq 1 \tag{2}$$

To proceed with our derivation, write the equations for λ for the two strategies as follows, by separating the terms for age a and substituting from equations (1) and (2):

$$1 = \lambda_1^{-a} l_1(a) m_1(a) + \sum_{a+1}^{\infty} \lambda_1^{-x} l_1(x) m_1(x) \tag{3}$$

$$1 = SM \sum_{a+1}^{\infty} \lambda_2^{-x} l_1(x) m_1(x) \tag{4}$$

Subscripts indicate strategies 1 and 2. Note that under these assumptions reproduction at age a has no effect on annual mortality in any year other than between a and $a + 1$, while reproduction at age a can affect annual fecundity at all later ages. We consider below conditions under which delayed reproductive effort might increase survival ($S > 1$) during the year in which reproduction is deferred or increase fecundity ($M > 1$) in later years.

If we now ask for the conditions under which $\lambda_2 = \lambda_1$, the summations in equations (3) and (4) become equal. Substituting from equation (4) into equation (3), we have

$$\frac{SM - 1}{SM} = \lambda^{-a} l_1(a) m_1(a), \qquad \lambda_2 = \lambda_1 = \lambda$$

For the conditions under which λ_2 is greater than, or less than, λ_1

$$\frac{SM - 1}{SM} \gtreqqless \lambda_1^{-a} l_1(a) m_1(a), , \qquad \lambda_2 \gtreqqless \lambda_1$$

If strategy 1 yields a stable population, so that $\lambda_1 = 1$, then

$$\frac{SM - 1}{SM} \gtreqqless l_1(a) m_1(a), \qquad \lambda_2 \gtreqqless \lambda_1 = 1 \tag{5}$$

This equation relates the probability that an individual can replace itself in the population by reproducing at age a, $l(a)m(a)$, and the consequences of reproduction at age a on its future chances for reproduction, SM. When $\lambda_2 > \lambda_1$ for animals in a particular environment, a lineage of individuals that delay reproduction to age $a + 1$ will multiply faster than a lineage of individuals that begin reproduction at age a (see Appendix and Wittenberger, 1979).

All four parameters in equation (5) can be measured in natural populations; $l(a)$ and $m(a)$ are respectively survival to age a and fecundity at age a if an individual tries to breed. Determinations of S and M depend entirely on comparing two separate sets of individuals in a population, those beginning to breed at some age a and those that defer reproductive attempts until one year later. S is the proportionate change in survival of individuals that defer reproduction, the survival between ages a and $a + 1$ of individuals deferring reproduction divided by the survival of individuals with early reproduction. M

is the proportionate change in the eventual fecundity of individuals that defer reproduction, the fecundity at any age $\geq (a + 1)$ of individuals that begin reproducing at age $a + 1$ divided by that of individuals that begin reproducing at age a. Note that M is *not* the ratio of fecundity at age a to that at age $a + 1$; it is the ratio of fecundities of individuals with different life-history strategies but with the same age $\geq (a + 1)$. A number of studies of natural populations of animals have shown that individuals have lower fecundity in their first year of reproduction than at later ages. I know of none, however, that compares individuals of the same age after they had begun reproduction at different ages. Wooller and Coulson (1977) have made a start in this direction.

In an earlier publication, I derived in a similar way a slightly different expression for the conditions under which delayed reproduction would evolve (Wiley, 1974b). That derivation involved different assumptions about the differences between an early reproducing and a late reproducing strategy and resulted in a somewhat different set of conditions for the evolution of deferred breeding. In particular, my earlier derivation assumed that deferred reproduction resulted in a translation of the function for age-specific fecundity $m(x)$ to later ages, so that $m_2(x + 1) = Mm_1(x)$, where subscripts 1 and 2 indicate strategies that begin reproductive effort at ages a and $a + 1$, respectively. Under this assumption, if individuals beginning to reproduce at age a have lower fecundity in their first year than later in life, then individuals deferring reproduction would also have lower fecundity in their first year of breeding than later. The fecundity of a first-time breeder at age $a + 1$ might still be larger by a factor M than the fecundity of a first-time breeder at age a. Under these conditions, deferred reproductive effort in a stable population should evolve when

$$AM > 1$$

where A ("As" in the terminology of Wiley, 1974b) equals the survival between ages a and $a + 1$ of individuals that begin reproduction at age a. In both of these formulations, when individuals that defer reproduction have high survival between ages a and $a + 1$ and high fecundity at ages $\geq (a + 1)$, then lineages of individuals with these characteristics tend to spread at the expense of those in which individuals begin reproduction earlier.

Deferment of reproductive effort could plausibly increase an individual's survival. The time and energy spent in reproductive activities, such as maintenance of a territory and advertisement activities, could instead be devoted to feeding and hiding from predators. There are very few data available, however, to test this possibility directly. Recent information suggests that male sage grouse survive better between ages 1 and 2 than when older (Beck and Braun, 1978). A direct test of the theory, however, would require a comparison of the survival of males in relation to their reproductive effort at age 1.

Deferred reproductive effort might increase later fecundity in several ways. A clear case is provided by animals with indeterminant growth. Early reproduction entails a diversion of resources from growth with the consequence that the animal is smaller in size at all subsequent ages, and size in animals with indeterminant growth usually correlates with fecundity. In addition, social organization that provides the possibility of queuing for advantageous reproductive positions also raises the possibility that at least partial deferment of reproductive activity could result in increased fecundity at later ages.

To test the theory we need comparisons of individuals that begin reproductive effort at early ages, without taking their places in queues of social positions, and individuals that at least partially defer reproductive effort by participating in queues. Certainly the magnitude of a male's eventual fecundity is substantial in lek-forming species like the sage grouse (see Wiley, 1974a). In addition, in at least some birds that live in stable social groups with one reproductive pair, groups are substantially more successful in reproduction than individuals that breed as pairs without nonreproductive helpers (Rowley, 1965; Woolfenden, 1975; Brown, 1978; Rabenold, personal communication). In both cases, we lack sufficient information for a thorough evaluation of the evolution of deferred reproduction. The key data will come from comparisons of the life-history parameters of individuals that pursue different strategies of reproductive effort early in life.

In analyzing the evolution of deferred reproduction, we have focused on the optimal age for an individual to begin reproduction on the basis of the rate of propagation of genes influencing the onset of reproductive activity. Although equation (5) compares the advantages of starting reproduction at only two ages, a and $a + 1$, it can also serve to determine the optimal age for starting reproduction. Just compare successive ages in pairs until at some age the conditions for further delay in the onset of reproduction no longer obtain. This optimal age for the onset of breeding will depend on the age-specific values of l_a, m_a, S, and M, the expected survival and fecundity at any age a, and the consequences of reproduction at age a on future survival and fecundity.

This analysis has not considered the possibility that alternate strategies for reproductive maturation might coexist stably in the same population. Some individuals might begin reproduction at an early age by parasitizing older individuals at the head of a queue, for instance by clandestinely stealing copulations or laying eggs in the nests of older animals. These individuals would thus jump the queue. In stripe-backed wrens, for instance, occasional (<5%) female auxiliaries develop full brood patches and thus could possibly have produced eggs. An occasional male auxiliary develops an unusually close association with the principal female of its group and might possibly copulate with her. In a diversity of species in which males defend locations at which they copulate with females, some males do not display and clandestinely steal copulations with females inside the areas defended by displaying males (Hogan-Warburg, 1966; van Rhijn, 1973; LeBoeuf, 1974; Constantz, 1975; Wells, 1977; Perrill et al., 1978; Howard, 1978).

Alternate strategies in the same population might constitute an evolutionarily stable mixed strategy (mixed ESS; Maynard Smith, 1974, 1976). When the successes of individuals using each strategy depend on their relative frequencies in the population, equilibrium frequencies could exist that would result in equal success for both strategies. Either individuals could change from one strategy to the other, with the proportions of time spent using each corresponding to the equilibrium frequencies, or individuals could each use one strategy, with the proportions of the different kinds of individuals in the population corresponding to the equilibrium frequencies. If the ages of individuals using the two strategies had different distributions, an analysis of the fitnesses of genes influencing the two strategies would have to consider fecundity and survival as functions of age for individuals with each strategy.

When individuals change their behavior progressively with age, the behavioral differences among individuals in a population might instead represent age-specific manifestations of a single optimal life history. In particular, when younger individuals can occasionally parasitize the reproductive success of older individuals without much effort or risk, they should do so, even when an optimal life history involves a deferment of more intense reproductive activity to later ages. If risk and effort are low, then the consequences of opportunistic reproduction for future survival and fecundity are likely to be small, and $(SM - 1)/SM$ in equation (5) will approach 0. Individuals should never pass up opportunities for reproduction that costs very little. When young individuals at the tail of a queue occasionally parasitize the reproductive advantages of older individuals at the head of a queue, the possibility thus exists that each individual pursues an optimal age-specific strategy.

VI. CONCLUSION

This article began with a suggestion for describing the movements of individuals through social positions in terms of ontogenetic trajectories. This approach led to several fundamental questions about the regulation of the ontogenetic trajectories of individuals in a society. Even the very little that we can now conclude about the mechanisms controlling ontogenetic trajectories led in turn to some fundamental evolutionary questions. Although the recent interest in animal societies has concentrated on the evolution of social behavior, nevertheless our perspective on ontogenetic trajectories as a basic feature of societies has uncovered some neglected evolutionary questions. A more thorough investigation of ontogenetic trajectories in animal societies, the behavioral and physiological mechanisms that control them, and their evolutionary adaptedness promises to open some new chapters in our understanding of societies as interacting individuals.

VII. ACKNOWLEDGMENTS

I thank K. Rabenold, C. Christensen, and M. Wiley for sharing their enthusiasm in the field and their information about stripe-backed wrens. Research discussed here was supported by the National Institute of Mental Health (MH 22316), the National Science Foundation (BNS 77-00279), and the Research Council of the University of North Carolina at Chapel Hill. This article is a contribution from the Behavioral Research Station in the North Carolina Botanical Garden. J. Russell and M. McVey offered valuable comments on the manuscript.

VIII. APPENDIX

Wittenberger (1979) has also obtained an expression for the conditions that favor the evolution of deferred reproduction. His formula is identical to my equation (5) with $M = 1$, although some algebra is necessary to see this identity. His condition for the evolution of deferred reproduction in a stable population is

$$\frac{b(\alpha - 1)}{b(\alpha)} < \frac{\Delta s}{1 - s(\alpha)} \tag{A-1}$$

where $b(\alpha - 1)$ is the fecundity of an individual breeding at age $x = \alpha - 1$, $b(\alpha)$ is the fecundity of all individuals at ages $x \geq \alpha$ regardless of earlier breeding, $s(\alpha)$ is the annual mortality of individuals at ages $x \geq \alpha$, and

$$\Delta s = s(\alpha - 1) - s'(\alpha - 1)$$

the difference in annual survival between individuals that breed at age $x = \alpha - 1$ (with prime) and individuals that defer reproduction to age

$x = \alpha$. Rearranging this expression, we obtain

$$b(\alpha - 1) < \Delta s \cdot b(\alpha) \sum_0^\infty s(\alpha)^z$$

where $z = x - \alpha$. Note that in a stable population

$$b(\alpha) l(\alpha) \sum_0^\infty s(\alpha)^z = 1$$

where $l(\alpha)$ is the survival from age 0 to age α.
Consequently Wittenberger's formula reduces to

$$b(\alpha - 1) < \Delta s / l(\alpha) \tag{A-2}$$

Now consider my equation (5) with $M = 1$. My S equals, in Wittenberger's terminology, $s(\alpha-1)/s'(\alpha-1)$, and my $l(\alpha)$ equals $l(\alpha)/s(\alpha-1)$. Thus (5) with $M = 1$ reduces to (A-2) above.

Equations (A-1) and (A-2) and equation (5) with $M = 1$ are equivalent conditions for the evolution of deferred reproduction in a stable population, when early reproduction has no effect on an individual's later fecundity. Wittenberger's application of equation (5) to males of polygynous species follows the traditional explanation for the evolution of polygyny in contrast to my two-factor theory.

Although Wittenberger (1979, p. 439) states that my earlier derivation (Wiley, 1974b) assumes constant fecundity throughout life for each strategy, in fact $m(x)$ in my derivations can take any form whatever. In addition, my derivations by incorporating the factor M include the possibility that early reproduction can influence later fecundity.

IX. REFERENCES

Allee, W. C. (1942). Group organization among vertebrates. *Science* **95**:289–293.
Beck, T. D. I., and Braun, C. E. (1978). Weights of Colorado sage grouse. *Condor* **80**:241–243.
Brian, M. V. (1979). Caste differentiation and division of labor. In: Hermann, H. R. (ed.), *Social Insects*, Vol. 1, Academic, New York, pp. 121–222.
Bronson, F. H. (1974). Pheromonal influences on reproductive activities in rodents. In: Birch, M. C. (ed.), *Pheromones*, Elsevier, New York, pp. 344–365.
Brown, J. C. (1978). Avian communal breeding systems. *Ann. Rev. Ecol. Syst.* **9**:123–155.
Canon, W. B. (1932). *The Wisdom of the Body*, Norton, New York.
Charlesworth, B. (1973). Selection in populations with overlapping generations: V. Natural selection and life histories. *Am. Nat.* **107**:303–311.
Clements, F. E. (1935). Social origins and processes among plants. In: Murchison, C. (ed.), *A Handbook of Social Psychology*, Clark University Press, Worcester, pp. 22–48.
Comte, A. (1830). *Cours de Philosophie Positive*.
Constantz, G. D. (1975). Behavioral ecology of mating in the Gila topminnow, *Poeciliopsis occidentalis* (Cyprinodontiformes: Poeciliidae). *Ecology* **56**:966–973.

Dementiev, G. P., and Gladkov, N. A. (1967). *Birds of the Soviet Union*, Vol. 6, Israel Program for Scientific Translations, Jerusalem.

Durkheim, E. (1972). *Selected Writings*, Giddens, A. (ed.), Cambridge University Press, Cambridge.

Emerson, A. E. (1939). Social coordination and the super-organism. *Am. Midl. Nat.* **21**:182–209.

Emlen, S. T. (1978). Cooperative breeding. In: Krebs, J. R., and Davies, N. B. (eds.), *Behavioral Ecology*, Blackwell, Oxford, pp. 245–281.

Eng. R. L. (1963). Observations on the breeding biology of male sage grouse. *J. Wildl. Mngm.* **27**:841–846.

Frame, L. H., and Frame, G. W. (1976). Female African wild dogs emigrate. *Nature* **263**:227–229.

Gadgil, M., and Bossert, W. H. (1970). Life historical consequences of natural selection. *Am. Nat.* **104**:1–24.

Gaston, A. J. (1978). The evolution of group territorial behavior and cooperative breeding. *Am. Nat.* **112**:1091–1100.

Glass, P. (1960). Factors governing density in the chaffinch (*Fringilla coelebs*) in different types of wood. *Arch. Neerl. Zool.* **13**:466–472.

Gosling, L. M. (1974). The social behavior of Coke's hartebeest (*Alcelaphus buselaphus cokei*). In: Geist, V., and Walther, F. (eds.), *The Behavior of Ungulates and Its Relation to Management*. International Union for Conservation of Nature and Natural Resources, Morges, Switzerland, pp. 488–511.

Haldane, J. B. S. (1932). *The Causes of Evolution*. Longmans, London.

Hausfater, G. (1975). Dominance and reproduction in baboons (*Papio cynocephalus*): A quantitative analysis. In: *Contributions in Primatology*, Vol. 7, Karger, Basel.

Hogan-Warburg, A. J. (1966). Social behavior of the ruff, *Philomachus pugnax* (L). *Ardea* **54**:109–229.

Horn, H. S. (1978). Optimal tactics of reproduction and life-history. In: Krebs, J. R., and Davies, N. B. (eds.), *Behavioral Ecology*, Blackwell, Oxford, pp. 411–429.

Howard, R. D. (1978). The evolution of mating strategies in bullfrogs, *Rana catesbeiana*. *Evolution* **32**:850–871.

Jarman, P. J., and Jarman, M. V. (1974). Impala behavior and its relevance to management. In: Geist, V., and Walther, F. (eds.), *The Behavior of Ungulates and Its Relation to Management*. International Union for Conservation of Nature and Natural Resources, Morges, Switzerland, pp. 871–881.

Jennings, H. S. (1942). The transition from the individual to the social level. *Biol. Symp.* **8**:105–109.

Krebs, J. R. (1971). Territory and breeding density in the great tit. *Parus major* L. *Ecology* **52**:1–22.

Kruijt, J. P., de Vos, G. J., and Bossema, I. (1972). The arena system of black grouse. *Proceedings of the XV International Ornithological Congress*. pp. 399–423.

LeBoeuf, B. J. (1974). Male-male competition and reproductive success in elephant seals. *Am. Zool.* **14**:163–176.

Lindauer, M. (1952). Ein Beitrag zur Frage der Arbeitsteilung im Bienenstaat. *Z. Vergl. Physiol.* **34**:299–345.

Maynard Smith, J. (1974). The theory of games and the evolution of animal conflict. *J. Theoret. Biol.* **47**:209–221.

Maynard Smith, J. (1976). Evolution and the theory of games. *Am. Sci.* **64**:41–45.

Maynard Smith, J. (1977). Parental investment—a prospective analysis. *Anim. Behav.* **25**:1–9.

Maynard Smith, J. (1978). *The Evolution of Sex*, Cambridge University Press, Cambridge.

McMahan, E. A. (1979). Temporal polyethism in termites. *Sociobiology* 4:153–168.

Mead, G. H. (1934). *Mind, Self and Society*. Morris, C. W. (ed.). University of Chicago Press, Chicago.

Mech, L. D. (1970). *The Wolf*, Natural History Press, New York.

Merton, R. K. (1949). *Social Theory and Social Structure*, Free Press, New York.

Miller, E. M. (1969). Caste differentiation in the lower termites. In: Krishna, K., and Weesner, F. M. (eds.), *Biology of Termites*, Vol. 1, Academic, New York, pp. 283–310.

Mills, J. S. (1843). *A System of Logic Ratiocinative and Inductive*, Longmans, London.

Noirot, C. (1969). Formation of castes in the higher termites. In Krishna, K., and Weesner, F. M. (eds.), *Biology of Termites*, Vol. 1, Academic, New York, pp. 311–350.

Orians, G. H. (1961). The ecology of blackbird (*Agelaius*) social systems. *Ecol. Monogr.* 31: 285–312.

Orians, G. H. (1969). On the evolution of mating systems in birds and mammals. *Am. Nat.* 103: 589–603.

Packer, C. (1979). Male dominance and reproductive activity in *Papio anubis*. *Anim. Behav.* 27: 37–45.

Parsons, T. (1951). *The Social System*, Free Press, New York.

Payne, R. B. (1969). Breeding seasons and reproductive physiology of tricolored blackbirds and redwinged blackbirds. *Univ. Calif. Publ. Zool.* 90:1–137.

Perrill, S. A., Gerhardt, H. C., and Daniel, R. (1978). Sexual parasitism in the green tree frog (*Hyla cinerea*). *Science* 200:1179–1180.

Rabenold, K. N., and Christensen, C. R. (1979). Effects of aggregation on feeding and survival in a communal wren. *Behav. Ecol. Sociobiol.* 6:39–44.

Rasa, O. A. E. (1977). The ethology and sociology of the dwarf mongoose (*Helogale undulata rufula*). *Z. Tierpsychol.* 43:337–406.

Rippin, A. B., and Boag, D. A. (1974). Spatial organization among male sharp-tailed grouse on arenas. *Can. J. Zool.* 52:591–597.

Rowley, I. (1965). The life history of the superb blue wren *Malurus cyaneus*. *Emu* 64:251–297.

Sarbin, T. R., and Allen, V. L. (1969). Role theory. In: Lindzey, G., and Aronson, E. (eds.), *The Handbook of Social Psychology*, 2nd ed., Vol. 1, Addison-Wesley, Reading, Mass., pp. 488–567.

Schneirla, T. C. (1971). *Army Ants: A Study in Social Organization*. Topoff, H. R. (ed.), Freeman, San Francisco.

Searcy, W. A. (1979a). Male characteristics and pairing success in red-winged blackbirds. *Auk* 96:353–363.

Searcy, W. A. (1979b). Sexual selection and body size in male red-winged blackbirds. *Evolution* 33:649–661.

Selander, R. (1972). Sexual selection and dimorphism in birds. In: Campbell, B. (ed.), *Sexual Selection and the Descent of Man, 1871–1971*, Aldine, Chicago, pp. 180–230.

Selander, R. K., and Hauser, R. J. (1965). Gonadal and behavioral cycles in the great-tailed grackle. *Condor* 67:157–182.

Spinage, C. A. (1969). Territoriality and social organization of the Uganda defassa waterback *Kobus defassa ugandae* Neumann. *J. Zool. London* 159:329–361.

Stearns, S. C. (1976). Life-history tactics: a review of the ideas. *Q. Rev. Biol.* 51:3–47.

Vandenbergh, J. G. (1974). Social determinants of the onset of puberty in rodents. *J. Sex Res.* 10:181–193.

van Rhijn, J. G. (1973). Behavioral dimorphism in male ruffs, *Philomachus pugnax* (L.). *Behaviour* 47:10–229.

Verner, J. (1964). Evolution of polygamy in the long-billed marsh wren. *Evolution* 18:252–261.

Vos. G. J. de (1979). Adaptedness of arena behavior in black grouse (*Tetrao tetrix*) and other grouse species (Tetraonidae). *Behavior* 68:277–314.

Wade, T. D. (1978). Status and hierarchy in nonhuman primate societies. In: Bateson, P. P. G., and Klopfer, P. H. (eds.), *Perspectives in Ethology*, Vol. 3; Plenum, New York, pp. 109–134.

Watson, A., and Moss, R. (1972). A current model of population dynamics in red grouse. *Proceedings of the XV International Ornithological Congress*, pp. 134–149.

Wells, K. D. (1977). The social behavior of anuran amphibians. *Anim. Behav.* **25**:666–693.

Wheeler, W. M. (1926). Emergent evolution and the social. *Science* **64**:433–440.

Wiley, R. H. (1973). Territoriality and non-random mating in the sage grouse *Centrocercus urophasianus*. *Anim. Behav. Monogr.* **6**:87–169.

Wiley, R. H. (1974a). Evolution of social organization and life history patterns among grouse. *Q. Rev. Biol.* **49**:201–227.

Wiley, R. H. (1974b). Effects of delayed reproduction on survival, fecundity, and the rate of population increase. *Am. Nat.* **108**:705–709.

Wiley, R. H. (1978). Social groups in tropical wrens. *Am. Phil. Soc. Year Book* **1977**:214.

Wiley, R. H., and Hartnett, S. A. (1976). Effects of interactions with older males on behavior and reproductive development in first-year male red-winged blackbirds *Agelauis phoeniceus*. *J. Exp. Zool.* **196**:231–242.

Wiley, R. H., and Wiley, M. S. (1977). Recognition of neighbors' duets by stripe-backed wrens *Campylorhynchus nuchalis*. *Behaviour* **62**:10–34.

Wilson, E. O. (1971). *Insect Societies*, Harvard University Press, Cambridge, Mass.

Wittenberger, J. (1979). A model for delayed reproduction in iteroparous animals. *Am. Nat.* **114**:439–446.

Woolfenden, G. E. (1975). Florida scrub jay helpers at the nest. *Auk* **92**:1–15.

Woolfenden, G., and Fitzpatrick, J. W. (1978). The inheritance of territory in group-breeding birds. *Bioscience* **28**:104–108.

Wooller, R. D., and Coulson, J. C. (1977). Factors affecting the age of first breeding of the kittiwake *Rissa tridactyla*. *Ibis* **119**:339–349.

Wright, P. L., and Wright, M. H. (1944). The reproductive cycle of the male red-winged blackbird. *Condor* **46**:46–59.

Yasukawa, K. (1979). Territory establishment in red-winged blackbirds: importance of aggressive behavior and experience. *Condor* **81**:258–264.

Zahavi, A. (1974). Communal nesting by the Arabian babbler: a case of individual selection. *Ibis* **116**:84–87.

Chapter 6

ON A POSSIBLE RELATION BETWEEN CULTURAL TRANSMISSION AND GENETICAL EVOLUTION

J. E. R. Staddon

Departments of Psychology and Zoology
Duke University
Durham, North Carolina 27706

I. ABSTRACT

Individuals respond differently to environmental stresses, and often a sub-population will respond more adaptively than the rest. The analogy with experiments on genetic assimilation suggests that such individuals are likely to share the genes responsible for their adaptive response. If the environmental stress in a social species is the presence of an innovative animal, the adaptive response may be imitation of the innovator by a subpopulation of individuals sharing genes responsible for this capacity. Small increments in the fitness of many such individuals may add up to a substantial selection pressure for the underlying genes. Since learning and imitation are not unitary abilities, but in all probability depend on underlying "computational elements," analogous to the instruction set of a computer, "assimilative" selection of this sort may constitute a positive feedback process that continually favors genes that add to the population's store of such elements. This hypothesis is consistent with such puzzling features of human social behavior as the enormous amount of time devoted to social intercourse and the disproportion between human intelligence and the demands placed upon it in primitive cultures. It is also compatible with the evolution of intelligence and a large brain size in non-tool-using social species such as whales and dolphins.

135

II. INTRODUCTION

Human beings are very much cleverer than other animals, and became so with extreme rapidity. During the past three million years, the brain size of man's ancestors grew at an unparalleled rate (e.g., Jerison, 1973; Washburn, 1978). We must presume that this growth was accompanied by a corresponding increase in intelligence. During the past 75,000 years, cultural evolution has been precipitous. The reasons for these changes, their magnitude, and their rapidity are still not fully understood. In this chapter I suggest a possible mechanism by which cultural transmission might directly influence and greatly facilitate genetical evolution.

The rapid development of human intelligence in the absence of any obvious environmental challenge points to a selection process involving positive feedback (cf. Wilson, 1975). I will argue that the differentiated nature of intelligent behavior, the high percentage of shared genes in humans, and the ability of environmental change to act as a genetic "probe," after the fashion described in experiments on genetic assimilation, provide the ingredients for such an autocatalytic process.

III. COMPONENTS OF INTELLIGENCE

Intelligence was at one time thought of as a generalized capacity, like physical strength or the ability to hold one's breath. Like strength, it could be measured on a single scale, with some individuals possessing more of it than others. As a practical matter, this linear view has turned out to be useful: even mathematicians, poets, and biologists can be rank-ordered, and measured IQ often has something to do with such agreed-upon rankings. But as a theoretical concept, the linear view is questionable. Increasingly refined statistical techniques have made it possible to break down into a variety of subtraits even the rather restricted set of abilities reflected in IQ questionnaires (e.g., Guilford and Hoepfner, 1971).

Advances in the study of artificial intelligence tend to reinforce this more differentiated view, as well as suggesting why statistical comparison allows the extraction of a "general" factor. For example, consider two identical programs, written in some relatively high-level language such as BASIC and designed to handle the same range of tasks. The range can be narrow or wide. Let us suppose that the two programs are run on two similar computers, but that computer A (say) has a slight superiority in respect of a single machine-language instruction: its "branch" is executed a little faster than the comparable instruction for machine B, say. A primitive instruction like this is likely to be executed many times in almost any program on almost any task. Hence careful comparison of the overall performance of the two machines will certainly reveal an overall superiority for machine A. Yet this reflects not some generalized "ability," but rather a very specific superiority at a level remote from that of the tasks the machine is performing. In other words, very specific differences at the level of what might be termed the internal or "machine" language of the brain will be reflected as "general" traits at the level of overt behavior. Hence statistical evidence for a general trait is not in conflict with the idea that intelligence depends upon numerous rather specific computational elements.

Learning ability, the animal equivalent of human general intelligence, has also been dethroned from its position as a unitary trait. For example, careful study of the learning process with individual animals reveals not the smooth "strengthening" of the correct response required by classical theory, but rather *selection* by reward and punishment of a limited set of effective behaviors from an initial pool of highly variable behaviors (Staddon, 1973). This initial pool is not random. It is constrained by species and situation. For example, the appearance of a novel stimulus in a familiar situation produces in rats a regular sequence of relatively stereotyped "exploratory" behaviors (Kello, 1973). A rat subjected to electric shocks will either freeze, attack a conspecific if one is present, or flee if cover is available. If the shock is delivered according to an escape or avoidance schedule, the animal will learn only if the required avoidance response is not too different from these "natural" defensive reactions (cf. Bolles, 1970). Similarly, pigeons will easily learn to peck for food, but will learn to treadle-press only with difficulty and cannot learn to wingflap for food at all. Rats in a multiple-arm radial maze easily learn to choose a different arm on successive trials, even if all are baited on every trial. It takes them much longer to learn to return consistently to the same arm, even if only one is baited (Olton and Samuelson, 1976). There are numerous other examples (e.g., Staddon and Simmelhag, 1971; Shettleworth, 1972). Learning evidently consists in the putting together of elements, selected from a limited set, under the guidance of the positive or negative consequences of different combinations (cf. Young, 1964).

The evolution of learning and intelligence can perhaps be thought of as the accumulation of more, and more effective, computational elements. This is not a novel view. Compare, for example, William James' (1910) comment on the evolution of intelligence:

> The whole history of our dealings with the lower animals is the history of our taking advantage of the way in which they judge of everything by its mere label, as it were, so as to ensnare or kill them. But as [Nature's] children get higher, and their lives more precious, she ... implants contrary impulses to act on many classes of things, and

leaves it to slight alterations in the conditions of the individual case to decide which impulse shall carry the day. Thus, greediness and suspicion, curiosity and timidity, coyness and desire, bashfulness and vanity, sociability and pugnacity seem to shoot over into each other as quickly, and to remain in an unstable equilibrium, in the higher birds and mammals as in man. They are all impulses, congenital, blind at first, and productive of motor reactions of a rigorously determined sort. Each one of them, then, is an instinct, as instincts are commonly defined. But they contradict each other — "experience" in each particular opportunity of application usually deciding the issue. The animal that exhibits them loses the "instinctive" demeanor and appears to lead a life of hesitation and choice, an intellectual life; not, however, because he has no instincts — rather because he has so many that they block each other's path. (v. 2, pp. 392–393)

If we charitably pass over James's metaphorical allusions to a prospicient "Nature," the resemblance between his view of choice and modern accounts of decision-making in terms of competition and time sharing (e.g., Ludlow, 1976; McFarland 1974; Staddon, 1977) is remarkable. If for "instinct" we substitute "computational element," James's view of intelligence is precisely the one I am proposing.

We are far from understanding either what these elements are or how they combine to produce adaptive behavior, but these details are not critical for the present story.

IV. HABIT AND INSTINCT

There is a well known complementarity between learning and instinct. As we have seen, learning requires a pool of behavioral variation from which the elements necessary for the correct act can be selected. This pool changes somewhat with training, but obviously learning will be rapid only if the to-be-learned behavior is in the initial pool. Rapid learning implies that the target behavior is largely preformed; it is not a reflection of the speeding up of a nonexistent "strengthening" process. Rapidly learned behavior is already largely instinctive. Darwin's (1872) comment on Mozart's musical precocity says it well:

> . . . the resemblance between . . .a habit and an instinct [is] so close as not to be distinguished. If Mozart, instead of playing the pianoforte at three years old with wonderfully little practice, had played a tune with no practice at all, he might truly be said to have done so instinctively. (p. 267)

Song sparrows and numerous other birds sing elaborate songs with little or no experience. They fulfill Darwin's criterion for instinctive behavior. Many other song birds require some experience; but in all learning is selective in ways that emphasize the continuity between learning and instinct.

This continuity is related to the striking asymmetry between production and reproduction that is characteristic of all learning. In human behavior it is usually discussed as the asymmetry between *recognition* and *recall*: under all but the most exceptional conditions, it is easier to recognize a word or a face than to produce or describe it. The tip-of-the-tongue phenomenon is a familiar example: a piece of information cannot be recalled, yet it is recognized instantly. The constraints on imprinting are of a similar sort. Thus, the swamp sparrow cannot sing its proper song without the opportunity to hear a male conspecific. Yet it unerringly selects swamp sparrow song fragments from a mixed song with elements from both swamp and song sparrow songs (Marler and Peters, 1977). Given a choice between conspecifics and artifical stimuli that would be effective if presented alone, ducklings generally prefer their own kind. Only under rather specialized conditions — in recognition experiments with carefully chosen distractor stimuli (Watkins and Tulving, 1975), or when natural selection pressures have been asymmetrical (supernormal stimuli, Tinbergen, 1948; Staddon, 1975) — is recognition notably inaccurate or inferior to recall.

Imitation shows the same kind of asymmetry. Thus, a single macaque discovered how to use flotation in water to separate wheat grains from sand in the Japanese colony at Koshima, but many were able to learn from her (Kawamura, 1963; see also Lewin, 1976). It is likely that the opening of milk bottles by tits that spread through Europe a few yers ago was also initiated by one or a few individuals and then imitated by many.

Granted that rapidity of learning means that many of the to-be-learned elements are already present, it is reasonable to infer that the most successful imitators are those who, had circumstances been slightly different, might themselves have been discoverers of that which they are now imitating.

It is a small step from inferring phenotypic similarity between a discoverer and his imitators to postulating a degree of genetic similarity. In most species, including man, a substantial fraction of genes is shared (e.g., Ayala, 1978). Hence, it is by no means implausible to suppose that individuals only distantly related by descent may nevertheless share particular sets of genes responsible for some phenotypic character. Often, therefore, imitators and imitated may share those genes responsible for the original creative act.[1]

This hypothesis can be phrased in general terms as follows: Consider a particular, desired phenotypic character, such as the capacity to compose music or solve a particular class of problem. Let us suppose that a few in-

---

For ease of communication I present the argument in black-and-white terms, speaking of "responsible for" when I can only mean "partially responsible," "share" when I mean "have more relevant genes in common than a randomly selected pair," and so on. I leave the sympathetic reader to insert the necessary qualifications — and hope that the unsympathetic one will not dismiss the argument out of hand because it is not complete in every detail.

novative individuals can solve the problem with minimal environmental help, but others, a larger number, can do so with a modest degree of prompting. The hypothesis, then, is that these two groups are genetically closer (with respect to the genes underlying the ability) than otherwise similar but randomly selected individuals. Consequently, were we to attempt to select for this ability, we would do almost as well to begin with the prompted group as with the innovators themselves.

In this form, the argument does little more than restate the conclusions of Waddington's (1960) well-known genetic assimilation experiments. For example, in one experiment:

> ... heat shock was applied to [*Drosophila*] pupae of an age which was known to be suitable for producing a number of phenocopies affecting the cross-veins .... If selection was exercised for any specific one of these types of phenocopy, strains could be rapidly built up which responded to the standard stress by a high frequency of this particular developmental abnormality. Moreover, after fairly intensive selection it was possible to produce strains in which the particular modification which had been selected for appeared in high frequency *even in the absence of the stress.* (pp. 393–394, my italics)

To select for a trait it is first necessary to produce phenotypes showing it or some approximation to it. Waddington's results suggest that in the case of an all-or-none trait, a special environment may be necessary to yield the required phenotypes. Nevertheless, selection from individuals so produced is likely to be effective in eventually producing a population that will show the required phenotype in the absence of the special environment necessary at first.

Haldane (1964/1959) has pointed out how this process may underlie the evolution of an instinct:

> Suppose that in Area A a particular volatile substance is produced by a nutritious plant, in area B by a poisonous plant. In area A those insects of a certain species which learn most readily to recognize this odour and associate it with food are at an advantage. As the features of the nervous system which favor such learning are accentuated, a few insects appear to whom the odour is attractive without learning, as the odour of a sheep appears to be attractive to sheep-dog puppies. They are at a double advantage, and after some time all members of the insect species are attracted by the odour without any learning. Similarly, in area B a race evolves which finds the odour repulsive. (p. 146)

To the extent that intelligence is the accumulation of "instincts," as James believed, Haldane's argument can be applied to its evolution also, as I now argue.

V. ASSIMILATIVE SELECTION

It appears to have escaped general attention that genetic assimilation provides a mechanism for selection that is not limited to individuals closely related by descent. Any environmental change constitutes a genetic probe, analogous to the heat stress in Waddington's experiments or the association of odor and food value in Haldane's example. Phenotypes so produced— *Drosophila* without cross veins or insects that seek a particular odor— constitute a subpopulation that is genetically similar with respect to the genes responsible for the phenotypic response, although these individuals need not be closely related by descent.

In the present context, the environmental change of special interest is the occurrence of a discovery by some member of a primate group. The phenotypes produced by this environmental change are the altered behavior of successful imitators. The selection is the fitness benefit conferred by the new behavior.

With his usual prescience, Darwin (1874) anticipated much of this argument although, being ignorant of modern genetics, he used it as a possible basis for group selection:

> It deserves notice that as soon as the progenitors of man became social. . .the principle of imitation and reason and experience would have increased and much modified the intellectual powers in a way, of which we see only traces in the lower animals. . . .Now, if some one man in a tribe, more sagacious than the others, invented a new snare or weapon, or other means of attack or defense, the plainest self-interest, without the assistance of much reasoning power, would prompt the other members to imitate him; and all would thus profit. (p. 146)

In an anticipation of kin selection, he adds:

> "Even if [the inventors] left no children, the tribe would still include their blood relations; and it has been ascertained by agriculturists that by preserving and breeding from the family of an animal which when slaughtered was found to be valuable, the desired character has been obtained." (ibid., pp. 146–147)

The analogy with genetic assimilation implies that the fitness benefit conferred on imitators will favor the relevant genes of the innovator, even if innovator and imitators are not directly related. The genetic interests of innovator and imitators are different, however, in that the innovator receives more benefit if he communicates with relatives than with nonrelatives. If he is imitated by relatives, then genes in addition to those underlying the beneficial behavior are also benefitted; otherwise, only the "innovator-imitator" genes he shares with nonrelative imitators are benefitted.

VI. EVOLUTION OF INTELLIGENCE

Assimilative selection allows for the rapid dissemination of genes underlying an adaptive, novel behavior originated by one or a few individuals and imitated by many. However, the process is self-limiting with respect to any given behavior, since the relative benefit conferred on an individual by a new idea will, in general, be inversely related to the number of his competitors who also know it. Selection for innovation is *apostatic* (Clarke, 1962), since rarity is favored. However, it is also *stabilizing*, since inability to learn a beneficial, but no longer novel, activity will incur a fitness cost, even if successful learning occasions no significant gain over one's competitors. In this sense, apostatic behavioral selection is different from the apostatic selection that leads to color polymorphisms, for example. In the latter case, there is no fitness loss associated with loss of an old color as a new one is acquired; but in the behavioral case, animals that trade one specific learning ability for another receive little benefit, whereas the acquisition of a new talent without loss of the old may give considerable advantage.

An appropriate analogy is perhaps to the evolution of cryptic coloration, where each successive approximation to the background color and pattern is doubly favored: first by providing better camouflage, and second by causing the phenotype to deviate from current predator search images. Moreover, the rarity benefit associated with loss of a widespread cryptic feature is unlikely to compensate for the reduction in crypticity itself. Similarly, inability to learn a widespread (hence beneficial) activity is unlikely to be compensated for by a unique ability to do some totally new thing.

The joint action of these two processes—apostatic, assimilative selection for genes underlying novel behavior; stabilizing selection for the capacity to imitate preexisting modes of behavior—implies a population of individuals capable of diverse modes of behavior, behaviorally polymorphic individuals, rather than a polymorphic population. The earlier argument suggested that intelligence represents the accumulation of genetically based computational elements. Clearly, the joint selection process just described will especially favor elements useful in more than one kind of behavior. Hence, selection for diverse behavior will at the same time be selection for an "instruction set" of maximal generality, that is, for intelligence.

Assimilative selection of innovator-imitator genes will be greatly facilitated by improvement in means of communication. In a prelinguistic species, imitation is confined to the immediate vicinity of the model. While this model allows a much more rapid spread of his influence than genetic transmission by descent, it is much slower than the dissemination permitted by spoken and written language. A prelinguistic innovator may influence tens or hundreds of other individuals in a year. Spoken language increases this number to thousands, and written (or electronic) communication expands it to millions.

This hypothesis suggests that human genetical evolution, far from having ceased with the advent of language and culture, may actually have speeded up (cf. Jaynes, 1978). The view also explains why primitive hunter-gatherer peoples, which, contrary to earlier ideas, have a great deal of leisure time (e.g., Pfeiffer, 1969; Sahlins, 1974), spend most of it in conversation. It accounts for the fascination of modern media communication, for widespread addiction to reading and television; for the power of the "Eureka!" experience and the almost irresistible urge to communicate it; and for the evolutionary paradox of the often meager direct fitness benefits conferred by invention upon the inventors themselves. These things are hardly explicable by the minute fitness benefit conferred on the individuals who learn of new things (cf. Humphrey, 1976). They are more readily explicable by the cumulative aggregate fitness increment conferred on the distributed genes underlying the learning.

The same process may also account for the evolution of a high degree of intelligence in the cetaceans, a group that shares with mankind almost none of the features often thought to be crucial to the evolution of human intelligence. Cetaceans are not bipedal, they use no tools, and they live in an undemanding and relatively impoverished environment. The one characteristic that the *Delphinidae*, particularly, share with man is that they are highly social and imitative (cf. review in Wilson, 1975). Their large brain and associated intelligence may alike result from the joint selection processes I have described.

Dawkins (1976) has engagingly argued for a process of cultural evolution via the dissemination of intellectual units that he terms *memes*. The major omission in his account is some independent process of selection. Memes spread, some more than others, but Dawkins points to no mechanism analogous to reproductive success to account for why some spread more than others. Without such a process, his account is circular: those that spread are more successful; they are successful because they spread. The present argument traces the selection of memes, like the selection of their genetic basis, back again to the fitness benefit conferred on individual organisms. Successful memes spread because they make use of computational units that have proven useful to their bearers in the history of the race. A memorable musical phrase (say) may be widely imitated because elements sensitive to symmetry, pattern, and rhythm are involved in many activities of evolutionary importance, such as speech, motor coordination and perception.

Both man and the higher primates appear to be much more intelligent than their environment demands. The present account sheds light on this phenomenon in two ways. First, assimilative selection acts on the ingredients of intelligent behavior, not the behavior itself. Even so simple a device as a hand calculator can easily be manipulated to do things never envisaged by its designers. It is not far-fetched, therefore, to suppose that the ingredients of intelligence may have evolved in the service of behaviors much simpler than those their possessors now (on occasion) display. Second, once the trend toward intelligence has begun, the rules of the game begin to change. Intraspecific competition — for mates, for resources — becomes more subtle: the smarter one's opponent, the more room there is for maneuver and strategy. A child may beat an adult at tic-tac-toe or even checkers, but chess or politics provides more scope for adult intelligence and the older player has the advantage. Hence, each increment in the general intelligence of the human population provides a social environment in which further increments are favored (cf. Jolly, 1966).

VII. ACKNOWLEDGMENTS

I thank Pat Bateson and Nick Humphrey for penetrating comments on an earlier version. I am especially grateful to Janis Antonovics, who managed not to sneer while subjecting a previous draft to near-fatal selection pressure.

VIII. REFERENCES

Ayala, F. J. (1978). The mechanisms of evolution. *Sci. Am.* **239**:56—69.
Bolles, R. C. (1970). Species-specific defense reactions and avoidance learning. *Psychol. Rev.* **77**:32—48.
Clarke, B. C. (1962). Balanced polymorphism and the diversity of sympatric species. *Sys. Assoc. Publ.* **4**:47—70.
Darwin, C. (1951). *The Origin of Species.* Oxford University Press, Oxford (Reprinted from the sixth edition, 1872).
Darwin, C. (1874). *The Descent of Man and Selection in Relation to Sex*, A. L. Burt, New York
Dawkins, R. (1976). *The Selfish Gene.* Oxford University Press, New York.
Guilford, J. P. and Hoepfner, R. (1971). *The Analysis of Intelligence.* McGraw-Hill, New York.
Haldane, J. B. S. (1964). Natural selection. In: Bell P. R. (ed.), *Darwin's Biological Work.* Wiley, New York (Reprint of 1959 Cambridge edition).
Humphrey, N. K. (1976). The social function of intellect. In: Bateson, P. P. G. and Hinde, R.A. (eds.), *Growing Points in Ethology.* Cambridge University Press, Cambridge.
James, W. (1910). *The Principles of Psychology.* Macmillan, London.
Jaynes, J. (1978). *The Origins of Consciousness in the Breakdown of the Bicameral Mind.* Houghton Mifflin, Boston.
Jerison, H. J. (1973). *Evolution of the Brain and Intelligence.* Academic Press, New York.
Jolly, A. (1966). Lemur social behavior and primate intelligence. *Science* **153**:501—506.
Kawamura, S (1963). The process of sub-culture propagation among Japanese macaques. In: Southwick, C. H. (ed.), *Primate Social Behavior: An Enduring Problem.* Van Nostrand, Princeton.

Kello, J. E. (1973). Observation of the behavior of rats running to reward and non-reward in an alleyway. Unpublished doctoral dissertation, Duke University.

Lewin, R. (1976). The rise of hunting in the pumphouse gang. *New Scientist*, October 7, 26–28.

Ludlow, A. R. (1976). The behaviour of a model animal. *Behaviour* **58**:131–172.

Marler, P., and Peters, S. (1977). Selective vocal learning in a sparrow. *Science* **198**:519–521.

McFárland, D. J. (1974). Time-sharing as a behavioral phenomenon. In: Lehrman, D., Hinde, R., and Shaw, E. (eds.), *Advances in the Study of Behavior*, Vol. 5, Academic Press, New York.

Olton, D. S., and Samuelson, R. J. (1976). Remembrance of places passed: Spatial memory in rats. *J. Exp. Psychol. Anim. Behav. Proc.* **2**:97–116.

Pfeiffer, J. E. (1969). *The Emergence of Man*. Harper and Row, New York.

Sahlins, M. (1974). *Stone Age Economics*. Tavistock Publications, London.

Shettleworth, S. J. (1972). Constraints on learning. In: Lehrman, D. S., Hinde, R. A., and Shaw, E. (eds.), *Advances in the Study of Behavior*, Vol. IV, Academic Press, New York.

Staddon, J. E. R. (1973). On the notion of cause, with applications to behaviorism. *Behaviorism* **1**:25–63.

Staddon, J. E. R. (1975). A note on the evolutionary significance of "supernormal" stimuli. *Am. Nat.* **109**:541–545.

Staddon, J. E. R. (1977). Schedule-induced behavior. In: Honig, W. K., and Staddon, J. E. R. (eds.), *Handbook of Operant Behavior*. Prentice-Hall, Englewood Cliffs.

Staddon, J. E. R., and Simmelhag, V. L. (1971). The "superstition" experiment: A re-examination of its implications for the principles of adaptive behavior. *Psychol. Rev.* **78**:3–43.

Tinbergen, N. (1948). Social releasers and the experimental method required for their study. *Wilson Bull.* **60**:6–51.

Waddington, C. H. (1960). Evolutionary adaptation. In: Tax, S. (ed.), *Evolution After Darwin*, Vol. 1, University of Chicago Press, Chicago.

Washburn, S. L. (1978). The evolution of man. *Sci. Am.* **239**:194–208.

Watkins, M. J., and Tulving, E. (1975). Episodic memory: When recognition fails. *J. Exp. Psychol.* **104**:5–29.

Wilson, E. O. (1975). *Sociobiology: The New Synthesis*. Harvard University Press, Cambridge, Mass.

Young, J. Z. (1964). *A Model of the Brain*. Clarendon Press, Oxford.

Chapter 7

THE BEHAVIOR OF ORGANISMS, AS IT IS LINKED TO GENES AND POPULATIONS

Wolfgang M. Schleidt

Department of Zoology
University of Maryland
College Park, Maryland 20742

I. ABSTRACT

Accepting Novikoff's notion of integrative levels (IL) in biology, we find that ethology is concerned predominantly with the IL of organisms. When we try to understand the evolutionary process, including the evolution of behavior, we must evoke, in addition to the IL of organisms, specific properties of the ILs of populations, ecosystems, and subcellular elements (genes). Because these four ILs are linked, any claim that evolution can be explained on the basis of principles solely at one level is ludicrous.

II. INTRODUCTION

Different vantage points give different perspectives and different perspectives lead to different observations. Differing observations of the same object lead to controversy which, at times, can increase the awareness of the scientific community. I doubt, however, that such controversy speeds up the understanding of a problem. I find it preferable to seek a variety of perspectives which may combine to a "hologram" of the problem.

The evolution of behavior has been viewed from various angles (Lorenz 1941, 1965, 1970, 1971; Wilson 1975; Dawkins 1976; Freedman 1979), and an underlying attitude appears to prevail: "My view shows how it really is, and the others look at the wrong thing." The implied value judgment is never stated that bluntly, of course, but is more or less carefully covered by scientific arguments and by hints that "the others" are simply not aware of certain important facts of life. I feel that this defiance results from a reluctance to accept vantage points which differ from those with which one is familiar, and that the problem can be resolved if we look at it in a broader context.

III. INTEGRATIVE LEVELS IN BIOLOGY

Since 1945, when Novikoff reviewed the concept of integrative levels in biology, this schema of structuring the universe and the domains of scientific disciplines has been generally accepted. I assume that Odum (1959), through his text *Fundamentals of Ecology*, has led a generation of biologists to accept the concept of integrative levels of organization as a basic fact. The most convincing argument for the validity of this schema is the clear ascendance in the size of the objects of study, from the subatomic level of physics, the atomic level of chemistry, through the levels of biology (subcellular, cellular, tissue, organ, organ system, organism, population, community, ecosystem, biosphere) to astronomy. Related to a level are the laws which govern the units on each level (Novikoff, 1945):

> Knowledge of the laws of the lower level is necessary for a full understanding of the higher level; yet the unique properties of phenomena at the higher level can not be predicted, *a priori*, from the laws of the lower level. The laws describing the *unique* properties of each level are qualitatively distinct, and their discovery requires methods of research and analysis appropriate to the particular level. These laws express the new organizing relationships of elementary units to each other and to the unit system as a whole. (p. 209)

Subscribing to the concept of integrative levels at the onset should not oblige us to accept it uncritically, and we must be prepared to modify the concept, if the empirical observations indicate inconsistencies. Furthermore, the idea of levels has the disadvantage of sneaking in a common value judgment, namely "the higher the level, the better" — similar to the common polarity of lower and higher animals; lower and higher functions of the CNS; *nukes* and *aspiro ad astra*. It may also imply that each level is discrete (like the stories of a building), or that "something" can assume different values along a particular scale (like a watertable, or the ambient noise level). None of these connotations is intended, and in order to keep confusion to a bare minimum, I will use here the acronym IL, to stand for "integrative level in the sense of Novikoff."

The essential characteristics of a system of ILs is that a unit on any particular IL is composed of units of the next lower IL and constitutes a member of a population which is the unit of the next higher level. For example, an organism is composed of subcellular elements, strands of DNA, mitochondria, etc., and is a member of an ecosystem which is composed of a variety of different organisms, with several additional ILs in between. For a particular purpose, we may strip this system to its essentials and consider only genes, organisms, populations, and ecosystems. The canon of the concept of ILs thus becomes very clear, though such a narrowing of view can lead us to overlook the importance of elements we have just excluded (e.g., the role of mitochondria, or of organ systems).

For a student of any particular IL, this scheme expressing the relations between different ILs implies that he must be familiar with the laws of the lower ILs, but is not required to know much about the higher ones. A chemist, for instance, must know the laws of physics, but can dispense with the laws of Mendelian segregation of traits. It follows that the laws of the lower levels will have, inherently, a much wider range of applicability, which leads to the intuitive conviction that they are more important. In combination with a preference for the simpler explanation of two ("Occam's razor"), this leads to a strategy of research which seeks to explain a phenomenon on IL_n by laws of IL_{n-1} and lower, and fosters a reductionistic attitude toward the problems of nature.

The relation between successive ILs can be more complex than Novikoff realized. This becomes apparent when we look at the linkage between any two consecutive ILs. This linkage is achieved by at least two types of relations: First, by the *inclusiveness of smaller units in larger units*, the parts making a whole, which is related to the overall trend of increase in size of the unit; and, second, by a degree of *persistence of the laws* which govern the relations among the units on any one IL and on the subsequent higher ILs. Newton's law of gravitation, for instance, applies to pebbles, apples, and celestial bodies. Within the domain of biology, at least, a third kind of linkage between ILs is provided by *feedback loops*. In other words, not only elements within the same level form an interdependent system (e.g., pituitary and ovary on the IL of organ systems), but elements on different ILs can be joined into *one* control system. For example, the reproductive state of an individual female rodent may be basically cyclical (a property on the IL organism), but also may be controlled by states of lower ILs (synthesis of steroid hormones, receptor sites in the hypothalamus, etc.) and on a higher IL (social status, state of the ecosystem). Now, it must be remembered that the interdependence of the elements within a control system which operates within its working range limits the applicability of causal linkage: blockage of the feedback loop at any place blocks the function of the whole system, not just the part downstream from the block. Only if we ignore the wider context of the system and zoom in on a small section of a feedback loop will we see cause and effect operating in a clear chain-of-command. Any linkage between ILs by feedback loops, therefore, invalidates the earlier mentioned implication of Novikoff's thesis — namely that *only the laws of a particular IL and those of the ILs below* are sufficient to fully understand a phenomenon.

I propose to amend Novikoff's thesis in this form: *Whenever ILs are linked by feedback loops, the laws of all ILs involved are applicable.* This defies the popular notion, noted above, that a scientist needs to know only the laws and principles governing the IL of his choice, and those below, and need not know any of those above (no more excuses for molecular biologists who plead ignorance of the laws of higher levels of biology!). Furthermore, if a particular phenomenon can be readily observed on one particular IL, we should not hastily conclude that this level is more important than others (as it appears in some questions raised about the level on which natural selection occurs, e.g., Lewontin, 1970; Alexander and Borgia, 1978), especially if an interaction among units on different ILs is obvious (e.g., Lewontin, 1974).

IV. BEHAVIOR: THE INTERACTION OF THE ORGANISM WITH ITS ENVIRONMENT

The study of behavior is clearly focused at the IL of organisms. This holds for the Skinnerian behaviorist as well as for the ethologist. There is considerable overlap with the adjacent ILs, of course, particularly in neurobiology, behavioral physiology, comparative ethology, and sociobiology. While other subdisciplines of biology, following the need for specialization, rarely look at the individual organisms as a whole, ethologists devote considerable time and effort to examining individual organisms in their natural environment, and have become the caretakers of the old legacy of natural history. Therefore, ethologists find it difficult to accept that it is possible to understand natural selection and evolution without considering the ongoing processes on the IL of organisms.

Organisms cannot exist without an environment; neither logically nor practically. In fact, we cannot think of evolution without an environment to which the organism can adapt. A few words about my view of the relation between the organism and its environment are therefore in order (for a more detailed discussion see Schleidt and Crawley, 1980). I propose to define the environment topographically as the space outside a particular organism's surface. This surface is not viewed as an impenetrable shield, nor as a single plane in the geometrical sense, but must be imagined as a *fuzzy zone* (fuzzy in the meaning of *fuzzy set theory* of Zadeh, 1965), with the physical and chemical properties of a complex semipermeable membrane. It is a filter zone which serves as a selective linkage between the individual and its environment. This linkage provides passageways or even active transport for matter and energy, and channels for information, especially for feedback loops between the organism and its environment (Ashby, 1956). If we freeze this surface zone in a momentary slice of time, we can detect none of the vital dynamic properties, but see only structure; we can describe only morphology. In the living organism, however, the surface is changing, however slightly, continuously; and it is this change of the surface structure over time which I propose to call *behavior* (Schleidt and Crawley, 1980). Such a definition of behavior subsumes not only the familiar topics of ethology, such as overt displays and change in location and orientation, but also changes in the dynamic equilibrium between organism and environment, such as desiccation of an animal in a dry atmosphere, or the change of our countryside by suburban sprawl.

When we observe an organism in its natural environment, we are amazed at the high degree of adaptation in structure and behavior, even though some individuals match the environment less well than others and consequently suffer reduced life expectancy and diminished reproductive success. The adaptedness of behavior cannot be explained without hypothesizing an interaction between the combined structural and behavioral characteristics of all interbreeding members of a population ("species-specific" morphology and behavior) with the characteristics of the collective environment of the individual members (species-specific niche). The environment is confronted, however, only by the individual on a *one-to-one relation*, as defined above by the linking surface zone, and *not by any population as such*. Since essential resources are either limited in their quantity, or are available only at a certain risk of encountering a predator, each individual influences the environment of the other members of its breeding population. Even if we restrict our interest to one aspect of behavior— its adaptedness— we find a complex interaction of elements at three ILs: organism, population, and ecosystem.

Similarly, if we trace the origins of overt behavior, starting within the surface zone, into the deep-structure of the individual organism, we find complex interactions in the whole gamut of ILs down to the molecular level, where sodium and calcium ions play their games on cell membranes.

V. INTEGRATIVE LEVELS IN THE EVOLUTIONARY PROCESS

In the early stages of evolutionary theories, it was assumed that natural selection acts on the individual organism. If one had sampled the opinion of biologists on the question: "Which level of organization within the realm of biology is most important for the understanding of evolution"? soon after the publication of Novikoff's paper, I suspect that a great majority would have pointed to the IL of organisms. Today, such an opinion-poll is likely to show that the choice is spread over the whole gamut of ILs of biology.[1] The importance of the IL of populations has been obvious to the geneticists for some time (especially through the work of R. A. Fisher), but it took the industrious effort of Wynne-Edwards (1962) and the bold formulations of Wilson (1975) to bring it to the general attention. Dawkins (1976), in his book *The Selfish Gene*, attempted to nail the problem precisely to the IL of genes. In his view organisms are nothing but weird *survival machines*, each controlled by the chairman of a committee of uncooperative, self-centered DNA-fragments; populations, therefore, are nothing but fleets of survival machines. Freedman (1979) has taken issue with such one-sided claims, and concluded that both organisms and populations are implicated. I feel that, in the last analysis, "everything is connected to everything" (as Richard Goldschmidt used to say), and that all levels from IL biosphere downward are involved in the evolutionary process. I suspect, however, that some connections contribute more than others. Therefore, I propose to isolate those ILs which are essential for the description of the evolutionary process, as it is believed to operate today in the animals commonly studied by ethologists. The qualifications are introduced as a convenience and for the sake of simplicity: I do not wish to become caught up in a lengthy discussion of the origins of life, or the evolution of viruses.

I propose to look at a generalized situation in which evolution is thought to occur: several individuals which potentially can interbreed, surrounded by their environment, confronting their limited resources and predators, ready to play the game of odds which we call *natural selection* (Fig. 1). Looking at the environment first, we note that it appears uniform in places, but overall it is best described as patchy, with more or less fuzzy borders between patches of finite resources and with clines of varying slope. As we focus on any one individual, we find that its environment most commonly includes a variety of other organisms, conspecifics (potential mates and/or competitors) and non-conspecifics (again potential competitors), as well as predators, and sources of matter, energy, and information. Even before making the first move in our game, we have evoked three of the ILs listed by Odum: organisms, populations, and ecosystems.

Let us assume we start our game in a "warm-up" round, during which more serious types of encounters among the participants are outlawed, and

---

[1] A straw vote among the faculty of our department gave the following result: organisms 36%; populations 18%; molecules, genes, cells, ecosystems 9% each; biosphere, universe 4.5% each ($N = 22$).

Fig. 1. Sets of escheroids in starting position for playing one round of "Natural Selection." The cast: *Escherus blobulatus* (center: pair ready for courtship, top right: mated pair, bottom right: pair with offspring); *E. neglectus* (bottom left: two individuals; note the lack of sexual dimorphism); *E. terribilis* (top left: one individual; preys on small escheroids and on juveniles of *E. b.* and *E. n.*); *Fraxinatoria dulcamare* (bottom and right edge: irregular patches of individuals; palatability and sex are unknown at this stage of the game); *Adalia escheri* (spread throughout the environment). Note that each individual faces a different local environment. Each individual is separated from its environment by a discrete surface zone, which is suited to display the phenotype of each individual, but at the same time covers the genes, which remain invisible throughout the game. The design for the escheroids was adapted from a woodcut by M. C. Escher (for details see Acknowledgments).

only playful behavior is permitted (no predation, aggression, reproduction). From the general morphology of each individual, and from its overt behavior, we can estimate its potential for success, like gamblers can gauge the potential success of a racehorse on the basis of its frame and its gait. However, still hidden from our view remains the potential for success which is encoded in the unique set of genes and in the unique bits of information stored during ontogeny [reflecting the phylogenetic adaptations to the ancestral environments, and the individually acquired adaptations of this organism to its own environment (Lorenz, 1965)]. It follows that we cannot describe the entities that are essential for the game of natural selection without considering a fourth IL of information storage at the IL of subcellular units of genetic information.[2]

Finally, the game has started, and our pieces move within the environment, reproduce, and expire, and new generations follow. Within the given ecosystem of our ballfield, equilibrium conditions will develop on each IL: alleles will arrange in a Hardy–Weinberg equilibrium, the number of offspring per pair will reach a stable distribution, and the social fabric will stabilize in a certain family structure or group size. Local disturbances will occur occasionally — a mutation in one gene locus, a juvenile male dropping off a cliff, a family dying after a meal of poisonous mushroom stew; but the probability for a lasting effect will be comparably small. A change in the environment which acts on many or all participants — a dramatic change in temperature (affecting the energy requirements) a change in topography (restricting individual mobility, and ultimately, gene flow), or the introduction of a new organism (affecting the ecosystem as a whole) — will have more spectacular effects. Now evolution comes in full swing, the game becomes exciting. The spectators lose their scientific objectivity and begin to cheer, revealing where their sympathies lie, and where they look: some stare at the ever-changing gene frequencies, others are fascinated by the hand-to-hand combat, and still others, high up in the ranks, find delight in seeing the cohorts move, gain, and wane. It appears as if three games were going on at the same time — a three-ring circus in one — genes playing against genes, individuals playing against individuals, and populations playing against populations. No single game can continue if any of the others is stopped — for the games are linked by the connections among the four levels of organization.

VI. ACKNOWLEDGMENTS

I thank my friends who helped me to understand several perspectives of the problem which originally were unfamiliar to me, and who improved the language of my manuscript, especially: Ellie and John Brown, Ed Buchler, Jackie Crawley, Nora Helgeson, Irene Magyar, Jane Potter, and Julie Riedel.

[2] Again for simplicity, I exclude other forms of information from further consideration. A more realistic model of evolution must provide for learning on the individual level, and for many species, including the human, it is imperative to incorporate the second channel for information transfer to successive generations, tradition.

The design for the spheres of the escheroids was adapted from a woodcut "Drie bollen I" (IX-45) by M. C. Escher (reproduced from *Graphik und Zeichnungen*, Heinz Moos Verlag, Munich, 1967, with the permission from the Escher Foundation — Haags Gemeentemuseum — The Hague). The escheroids are an invention and the sole property of the Author. Their reproduction, storage, or transmission in any form or by any means, photonic, electronic, magnetic, gravitational, chemical, biological, astronomical, or otherwise requires written permission from the Author.

VII. REFERENCES

Ashby, W. R. (1956). *An Introduction to Cybernetics*, Wiley, New York.
Alexander, R. D., and Borgia, G. (1978). Group selection, altruism, and the levels of organization of life. *Ann. Rev. Ecol. Syst.* **9**:449–474.
Dawkins, R. (1976). *The Selfish Gene*, Oxford University Press, Oxford.
Freedman, D. R. (1979). *Human Sociobiology*, The Free Press, New York.
Lewontin, R. C. (1970). The units of selection. *Ann. Rev. Ecol. Syst.* **1**:1–18.
Lewontin, R. C. (1974). *The Genetic Basis of Evolutionary Change*, Columbia University Press, New York.
Lorenz, K. (1941). Vergleichende Bewegungsstudien an Anatiden. *J. Ornitol.* **89**:194–293 (Engl. transl. in Lorenz 1971).
Lorenz, K. (1965). *Evolution and Modification of Behavior*, University of Chicago Press, Chicago.
Lorenz, K. (1970, 1971). *Studies in Animal and Human Behavior*, Harvard University Press, Cambridge, Massachusetts.
Novikoff, A. B. (1945). The concept of integrative levels and biology. *Science* **101**:209–215.
Odum, E. P. (1959). *Fundamendals of Ecology*, W. B. Saunders Company, Philadelphia.
Schleidt, W. M., and Crawley, J. N. (1980). Patterns in the behavior of organisms. *J. Social Biol. Struct.* **3**:1–15.
Wilson, E. O. (1975). *Sociobiology*, Harvard University Press, Cambridge, Massachusetts.
Wynne-Edwards, V. C. (1962). *Animal Dispersion in Relation to Social Behavior*, Oliver and Boyd, Edinburgh.
Zadeh, L. A. (1965). Fuzzy sets. *Inform. and Control* **8**:338–353.

Chapter 8

FROM CAUSATIONS TO TRANSLATIONS: WHAT BIOCHEMISTS CAN CONTRIBUTE TO THE STUDY OF BEHAVIOR

Steven P. R. Rose

Brain Research Group
Department of Biology
The Open University
Milton Keynes MK7 6AA, U.K.

I. ABSTRACT

This paper discusses the relationship between observations of phenomena at the behavioral level, and at the biochemical and physiological levels. It argues that a complete description of an organism and its behavior demands the integration of these levels, and it raises the question of the nature of such an integration. It begins by discussing the nature of "levels" of biological experimentation and explanation. It shows that purely "systems" explanations of behavioral observations cannot be complete, nor can they distinguish between alternative models. They also run the danger of reification, and the general tendency to reify explanations of behavior, especially among sociobiologists, is analyzed.

The fallacy of reductionist explanations of behavior, in which behavioral events are seen as "caused by" antecedent biochemical ones, is then discussed. Various conventional ways of springing the reductionist trap, especially those which see causes as running both "up" and "down" between levels, are shown to be inadequate or to lead to paradox, and the inadequacy of a purely correlational approach is described.

The alternative proposed herein is a nonreductive version of the identity hypothesis in which biochemical events are seen not as causing behavioral events but as synonymous with them, but in a different language system. It is proposed that the word "cause" can only be properly applied within a given level and that the relationships between epistemologically defined levels are best described as translational, as between different languages. Examples of how this mode of understanding may work in practice are given, and, finally, its relationship to the experimental work in the search for the biochemical events which occur during learning and memory formation, the storage of the "engram," are discussed.

II. ON LEVELS OF ANALYSIS

To ask a biochemist to contribute to a volume of *Perspectives in Ethology* must be a gamble for the editors; to agree to write it is to take a gamble of my own. Many people, firmly located in their own slice of the multidecked sandwich that constitutes the present state of the sciences of neurobiology and behavior, argue that the time is not really ripe to move even one slice up or down. While the ultimate aim should be an integrated account of brain and behavior at all levels from the biochemical and anatomical to the psychological and social, at the present time, they claim, we are best employed attemping to complete descriptions at our own levels of analysis; it will be for the future to fit the physiological layer onto the biochemical, and the psychological on top of that. There are also some behavioral scientists, I suspect, who are more than a little concerned that the "harder" disciplines may, given half a chance, engulf them whole — and indeed there has recently been a plethora of somewhat grandiloquent claims that such an engulfing is about to or even has already occurred. At least one aspect of the furor caused by the invention of the "new synthesis" of "sociobiology" has been about the scent-marking of disciplinary territories.

In this essay I want to argue that while the several brain and behavioral disciplines must continue to remain epistemologically distinct and that higher-level accounts of behavior are not going to disappear into the maw of the lower-level ones — now or in the future — each needs the other if it is to make sense of its own research questions.

I must pause for a moment at the outset and say a little more about the question of "levels." It is a commonly enough used term in the brain and behavioral sciences, yet it seems to me that its implications are often a bit taken for granted. In dealing with the sciences of complex systems like living organisms, it is generally assumed that there is a hierarchy of orders of analysis, from the physical to the social. Each level of the hierarchy corresponds approximately to the boundaries of one of the traditional scientific disciplines: physics, chemistry, biochemistry, physiology, psychology, etc. Levels are thus essentially operationally and epistemologically defined; clearly they are not ontologically distinct and their objects of study may be identical. Despite the fact that the rules of experiment and proof and the language of explanation, of each level, acquire a certain autonomy, the purposes of studying an object at one level may only become apparent in the context of the other levels at which it may be approached. (The significance of this perhaps slightly obscure formulation will, I hope, become apparent later.) Nonetheless, the hierarchy of levels is not considered symmetrical; it is given an upward and downward direction. "Upward" is seen as in the direction of increasing complexity; "downward" traditionally in the direction of increasing "fundamental" or "basic." But the levels also correspond approximately to the historical order in which the different scientific disciplines emerged, to the perceived "hardness" of the sciences, and, parenthetically, to the social ranking accorded their practitioners (on the supposed "primacy" of physics, see Levy Leblond, 1976).

Now it is clear that, at least at the present time, the levels of analysis in the hierarchy correspond to a social division of labor among researchers. Although there is some communication between levels, which is the subject of much of the discussion which follows, workers within each level strive toward a certain epistemological completeness within their own level, despite the more-or-less contested claims that in due course the totalized project of "the sciences as a whole"will succeed in integrating them in an as yet unspecified way. The debate is over the nature of this integration, and it ranges from hard-line reductionism to emergentism. Reductionism claims that, in the long run, higher-order levels will be collapsed into the lower-order ones; that they are "nothing but" particularly complex systems to the solution of which the equations of the physicist are slowly approaching [a good recent example of such a claim is the introduction by Wilson to *Sociobiology — The New Synthesis* (1975)]. Emergentism claims an ontological distinction between levels; the methods and laws of the lower-order disciplines are by their nature insufficient or inappropriate to deal with higher-order phenomena (e.g., Koestler and Smythies, 1969), or indeed that the higher orders, such as the phenomena of social existence and especially of the human condition, are largely or entirely divorced from the lower (e.g., if he can be taken at face value, Feyerabend, 1978, or much of the writings of structuralism, existential psychology, or psychoanalysis).

The problems of understanding the relationships between levels of analysis are not newly perceived (e.g., Engels, 1940; Needham, 1943) and the argument of this essay is derived from the tradition which claims that the relationships exist and that they must be sought in a materialist but non-reductionist, dialectical framework. It is not simply that as one ascends the hierarchy the integral objects of study of the lower levels become the compositional units of which the higher is perceived as being built up, but that new organizing relationships emerge between them and that an open system at one level becomes a closed component of a higher-level system. (For instance, physiology requires not merely the chemical and dynamic relationships of biochemistry; the spatial relationships between components, studied by anatomy and morphology, are also necessary.) It is in this sense that higher-order levels are more complex. The physiologist may study the properties of a single cell, the biochemist the averaged properties of a million such ground up with a homogenizer, and yet the complexity of the former, the order given by the relationship of its parts, is greater.

This is the "orthodox" solution within the dialectical tradition and I wish to add to it that the interpretation at each level requires concepts which are themselves only appropriate to that level; *genes* are not spiteful or altruistic, *assemblages of cells* cannot learn, love, or be angry; such terms are inappropriate to the genic or physiological levels of analysis, but appropriate to those of the whole organism.

It follows that there are important questions which require, if not the integration of biochemists and behavioral scientists, then at least that they should speak to one another. A task that we all face is the *identification* of particular phenomena as the subjects of analysis, the *description* of the phenomena in our own disciplinary languages, and the *translation* of these languages one into the other. In attempting this triple task we will find that the languages of all the disciplines become enriched and even transformed. Most of this essay is concerned with the conceptual and methodological problems that this task presents; toward the end I attempt to show, drawing on the experience of my own collaborative research with ethologists, physiologists, and anatomists over the past 14 years, how some of these problems may perhaps be approached. I should firmly emphasize that the discussion which precedes the account of the experimental research is my own viewpoint: it is not necessarily shared by Pat Bateson or Gabriel Horn, my collaborators in the imprinting experiments, or my colleagues within the Brain Research Group with whom many of the experiments have been done, and they must not be held responsible for it!

III. THE OBJECTS OF BEHAVIORAL STUDY

Perhaps one could begin by asking what a "complete" description of an organism might entail. I am using the word behavior to cover the questions which interest both ethologists and psychologists here and the ethological reader must pardon the naivete of someone looking at their discipline from the outside. Classically the "hard" sciences have claimed that they wish to be able to provide, for phenomena within their domain, an explanation and a prediction. Explanation takes any particular aspect or subset of the phenomena within the domain and accounts for it in the terms of the general laws or theorems of the domain and of particular antecedent causes. Causal explanations may be at one level of analysis of the phenomenon being studied, or they may be reductive between levels of analysis. A particular quantity of gas at a particular pressure will occupy a given volume; if the temperature is increased and the pressure kept constant the increase in volume can be approximately calculated by the application of a general equation; the volume increases *because* the temperature increases — at the reductive level because of the greater thermal energy imparted to the gas molecules. Reductive explanations and attempts to find reductive causes have been the dominant mode of science in the West since the days of Descartes.

The predictive aspects follow from explanation and reduction: given a particular quantity of gas, the temperature and pressure being known, one can predict the volume. However, the gas laws, which are explanatory and predictive for a given volume of gas, do not enable one to predict the motions of the individual molecules which compose the gas; it is the summed statistical properties of these molecules which are represented in the equation and there is an inherent impossibility about approaching the question of the properties of the mass of the gas by determining the velocity of each of the individual molecules of which it is composed and summing them. Indeed, there are potentially infinitely many possible combinations of velocities of particles which, when summed, would be represented by the formulation of a given mass of gas at given volume, pressure, and temperature, quite apart from the question of how to actually measure the velocity of any individual particle without in so doing interfering with it and hence changing its velocity.

One problem for any student of behavior is whether the objects of study are like a large mass of gas or like the individual particles of which it is composed. The answer is clearly that they are like both at the same time; a multitude of "internal" cellular states can, when summed, be represented by some observable piece of behavior on the part of the organism, and in this sense it behaves like a mass of gas composed of particles. At the same time the complexity of the components of which the organism is composed is vastly greater than that of the particles within the gas, subject to much more individual variation, so that individual differences occur between organisms which make their behavior explicable and predictive itself only in a statistical sense; thus the organism becomes more like a particle of gas than the summed mass of the gas, even before we insert the additional complexity of its interaction with other organisms. And there is a further crucial difference: both the gases and the particles of which they are composed are assumed to be neutral as to their history. Rules, explanations, and predictions appropriate and applicable to the study of gases are assumed to be unlimited in space and time. Organisms and populations of organisms have nonneutral histories whose logic is a vital part of their continuity; evolution for populations; development, individual experience, and social interaction for the organism impose singularities which are the very stuff of understanding them.

Indeed, the very use of the word "behavior" in this connection involves the dangers of impoverishing understanding, carrying over as it does a certain ideological baggage derived in part from its behaviorist past. For if behavior is defined only in terms of the observable actions and reactions of an organism to its environment, the entire internal dimension which the Watsonians and Skinnerians discarded is in danger of remaining irretrievable. Only the minimum of introspection is required for us to recognize that there may be a multitude of inner states of consciousness, intentionality, or "feeling" which may produce apparently identical outcomes in observable behavior, and we have no reason to assume that what is true for each of us as individuals is not true for other humans and, albeit in lesser measure, for nonhuman organisms with large brains and a rich behavioral repertoire (Rose, 1976). This question of "feeling" has been discussed by Heller (1979) and if I continue to use the word "behavior" here, I do so to include the domain of "inner states."

IV. THE INADEQUACY OF SYSTEMS APPROACHES

Clearly the study of "behavior" in this sense itself takes place at many levels. A population of interacting organisms can be studied both with a view to following the interactions of this population with the rest of the world and its evolutionary unfolding in time, and with a view to the relationships, successes, and failures of individuals within any given population (the domain now beginning to be colonized by population genetics and sociobiology). At this level, biochemistry has little to contribute, although its techniques may enable the molecular aspects of some types of interanimal communication to become clear (e.g., pheromones).

Even when the focus of study narrows to the individual organism, it can be considered both in its interaction with the outside world, including others of its own species, and as a product of its own developmental history. The systems approach to the study of the behavior of individual organisms (e.g., Oatley, 1978; Toates, 1980), although at present it largely ignores this developmental history, and necessarily largely isolates the individual from its social context, may indeed in due course be able to fulfill some of the goals of explanation and prediction at least a bit better than its intellectual precursor, behaviorism, without recourse to the cellular or biochemical actuality.

According to this view, organisms are seen as essentially rather sophisticated computers which must possess certain types of functional components in order to achieve specified outputs. The behavioral outcomes are specified in terms of the products of a number of rather arbitrary internal black boxes interconnected in specific manners. The task of research then becomes largely that of observing relevant behavior patterns and attempting to segment them into units (these for behaviorists used to be seen as particular patterns of motor activity; today they are more widely regarded as goal-directed functions that may employ a number of possible routes to a particular goal, be it food, orgasm, or whatever). The test for this approach is whether the explanatory black boxes can generate predictions as well as *post hoc* accounts; hence the enthusiasm for computer modeling.

Development along these lines of models which fulfill some of the canons of scientific theory would, I suspect, satisfy many ethologists, especially if plausible assumptions could be made that would encompass the phenomena of development and evolutionary origin — and indeed there is an increasing tendency for "evolutionary" explanations to be grafted onto such analyses simply by extending the black boxes downward. If there can be a box inside an organism labeled "comparator" or "glucose detector," why not let there be a *gene* labeled "spite" or "altruism" or whatever? So far as development is concerned one could look at the success of Piagetian models for genetic epistemology in interpreting the growth of cognition in humans, and the French enthusiasm for structuralism in general, as examples of modeling approaches which make no demands on a cellular biology or biochemistry for their explanations.

What, then, is the problem with this modeling approach by which the organism becomes essentially emptied out into abstract function boxes and flow charts? If it can generate explanations and prediction, is this not sufficient? Despite its popularity, modeling of this sort is in my view always flawed, being able to pass the theory test perhaps, but never the reality test.[1] As I pointed out in connection with the gas laws, an infinite number of combinations of a given number of particles with particular velocities can produce a particular volume of gas at a given pressure and temperature. Similarly, an infinite number of models is always possible for any given outcome being modeled; further, a test of a model which does not bear out its predictions can always be overcome by a modification of the model without

---

[1] Many modelers are, of course, well aware of this problem.

a change in its basic elements. After all, the Ptolemaic system of planets rolling in epicycles on the crystal globes of the heavens actually fitted the data as well, or better, than did the Copernican one; it just happens that the world is in reality not arranged as the Ptolomaics modeled it. There was no possible way of refuting the Ptolemaic model *merely* on the basis of better observations of the transit of Venus or the perihelion of Mercury; it had to be done by asking quite different sorts of questions about the way the planetary and stellar worlds were organized. This, incidentally, is the nub of one of the criticisms of sociobiology — that its search for models from games theory for animal behavior and adaptive strategies does not allow refutation. If the rules of the game are that you can postulate any number of arbitrary genes for behavioral processes, then any outcome can be "accounted" for in theoretical terms. The reality test asks: (a) Do the processes occur or are they arbitrary reifications? (b) Are the genes present? and (c) What is the relationship between the expression of particular genes in particular environments and the behavior patterns under study? It is here that systems-ethology and sociobiology come face-to-face with the real world, and must accommodate to it or become merely vacuous.

V. THE HAZARDS OF REIFICATION

A word about reification may be in order here. Science proceeds partly by attempting to identify similarities and differences. Thus it examines disparate phenomena and asks, first, are there common features which underlie and explain them all, and, second, what is different about the underlying features which then become expressed in surface differences (this is what a whole area of genetics is about, for instance). The problem is that when one extracts out of the blooming, buzzing confusion of the world around one some hypothetical common underlying feature, say x, this x then tends to take on a life of its own. Its reality, rather than hypothetical nature, becomes taken for granted as the paradigm for constructing future experiments; its use becomes extended as a way of explaining phenomena quite apart from those for which it was first hypothesized, and so on. Soon, x, from being a particular *ad hoc* extractive hypothesis, becomes the object in *terms of which* others must be explained. Think for instance of the way in which falling bodies are said to obey the law of gravity, as if the *law*, which is a theoretical construct, had primacy over the *objects* whose properties and whose relationships it is defining.

Much ethological and behavioral research proceeds in this way; x may be, e.g., perception, learning, hunger, or sexual need. The implication that there is a single underlying, reified thing, x, runs the danger of confusing thought instead of producing the looked-for clarifications, as with 19th-century phrenology (bumps of philoprogenitiveness, love of music, and so forth) or contemporary "altruism" or "aggression." Disparate phenomena may become quite inappropriately grouped in such classifications. Much of the history of experimental psychology can be seen in terms of tensions between "lumpers" and "splitters" — those who want to argue for unity among particular phenomena (for instance, verbal learning in humans and conditioned responses in experimental animals, or unitary "general intelligence factors" of the Spearman type) vs. those who emphasize distinctions (multiple forms of learning or Guilfordian multifactorial analyses).

There is an even greater hazard involved in reification, however, and that is when a dynamic process itself becomes turned into a thing, with something of the weight of a material object about it. Behavior patterns are dynamic interactions of the organism with its environment. One really should not speak of "altruism" or "intelligence" as *nouns* describing objects, but as *verbs* describing activities. They do not reside as static lumps inside the organism, to be drawn as a system-modelers box or reduced to mathematics as a population geneticist's phenotype. It is interesting that in a recent debate one prominent sociobiologist, Dawkins, came close to a recognition of this problem when discussing altruism. "Genes for altruism," he said, were any which conferred upon their possessor the property of acting "altruistically" in particular circumstances; thus "genes for bad teeth" in carnivores would result in their reducing their share of any common food resource to the benefit of others, and by definition this reduction was altruistic. Yet to define "bad-tooth genes" as "altruistic genes" is surely to further empty the reified abstraction of "altruism" of useful explanatory power. The analogy is with the conceptual difficulties that pre-Newtonian physics found itself in with the concept of force, until it could be seen that forces are not reified properties of material objects but the expression of relationships between them. Until ethology can distinguish the material objects within its domain from their relationships, the dangers of reification will continue to dog it.[2]

It is for this reason that psychology and ethology cannot ever complete their programs if they rely upon the arbitrary modeling of behavioral processes to account for and interpret observed aspects of behavior. Such interpretations must not only be in accord with the models, but in accord with the constituent physiology, anatomy, and biochemistry of the organism. No amount of modeling theoretical neuronal networks can be adequate until the biochemistry and morphology of the synapse is known; no amount of theorizing on the mode of genetic transmission is complete

[2] This criticism, of course, applies not merely to ethology; other sciences may be at least as guilty.

without knowing the structure of DNA. To those who argue that one can use a computer or drive a car without knowing the composition of the silicon chip or the physics of the internal combustion engine, I would reply that you may be able to use the machine but you could not design, build, or repair it; nor in fact could you produce a complete explanation of how it did what it did.

VI. THE REDUCTIONIST FALLACY

Now it is true that most biochemists do not at all see a need to impress ethologists with their own legitimacy; rather they take it for granted, and indeed are faintly surprised if anyone should tell them that it is not their manifest destiny in due course to imperialize the behavioral sciences as they believe they have done the physiological ones. To the biochemist, reductionism is not second but first nature. Not merely is there no question but that explanations must be bottom-up — DNA determines primary protein structure, determines the folding of protein chains and their enzymic activity, determines cellular architecture, determines function, determines organismic behavior — but further, for many there are not even multiple pathways, redundancies, or flexibilities to higher-order outcomes. "A disordered molecule causes a diseased mind" was the dictum of a respected biochemist at a conference on learning I attended in the U.S. in 1978.[3] Particular molecular events "cause" particular behavioral events. Just as a recessive mutation results in the production of a hemoglobin molecule with a particular single amino acid substitution and this in turn is able to alter the kinetics of oxygen binding to the hemoglobin molecule, "causing" the disease known as sickle cell anemia; and the phenylketonuric mutation "causes" an enzyme deficiency in the pathway of phenylalanine metabolism which "causes" irreversible mental retardation; so changed levels of GABA metabolism in the olfactory bulb of certain strains of mice "cause" aggressive behavior. Learning is "caused" by the synthesis of particular protein molecules in the cortex; changed dopamine levels "cause" schizophrenia in humans and the possession of an additional Y chromosome "causes" unusually violent males. If these events were *not* causal how does one explain the effects of drugs, or justify publishing academic journals with titles permuting the word psychopharmacology?

In this approach, it follows that the proper task of the sciences of the organism is the reduction of the individual's behavior to particular

[3] And indeed, what is this but a version of the notorious "Central Dogma" of Francis Crick: DNA→ RNA→protein writ large. When the history of the last 25 years of biology comes to be written, the confusion engendered by this dictum may rank as important as the golden helix itself.

molecular configurations, while the study of populations of organisms reduces to the search for DNA strands which code for reciprocal or selfish altruism. Paradigm cases of this approach over the last decade have been the attempts to purify RNA, protein, or peptide molecules that are produced by learning and which "code" for specific memories — for instance, the late Georges Ungar's scotophobin, the "fear of the dark" peptide (see Irwin, 1979; Rainbow, 1979; Rose and Langstaff, 1981, for reviews) or the molecular biologist's search for an organism with a "simple" nervous system that can be mapped by serial electron microscope sections and the different wiring diagrams associated with different behavioral mutations identified (e.g., Ward *et al.*, 1975). If the structuralist and modeling ethologists claim that everything lies in the program and the hardware is irrelevant, the Brenner/Ungar approaches certainly mistake the singer for the song.

The paradoxes that this type of reductionism gets itself into are probably more vicious than those of the systems modelers. The paradoxes have been apparent, of course, since Descartes, whose reduction of the organism to an animal machine powered by hydraulics had to be reconciled, for the human, with a free-willed soul in the pineal gland. As then, so today mechanistic reductionism forces itself into sheer idealism before it is through. In its modern form one finds it in molecular biologists whose ultimate ethic becomes the search for absolute scientific "truth" (Monod, 1973); physiologists, such as Eccles, for whom indeterminacy creeps in through the synaptic clefts (e.g., Popper and Eccles, 1977); and in sociobiologists who always rescue genetic reductionism in the final chapter by human consciousness and "our" free will galloping over the hill in the nick of time like the U.S. Cavalry. See, for example, the concluding paragraphs and chapters of, Dawkins (1976), with its "memes," or Barash (1976), and Wilson (1975), who allow that genetic determinism may, if "we" wish, be overturned by conscious human effort, the resurrection of an almost Spenglerian "will," for this ploy. For a beautifully clear account of the dilemma as it confronts the nondualist physiologist, see Wall's illuminating chapter "My Foot Hurts Me: The Analysis of a Sentence" (Wall, 1974).

VII. SPRINGING THE TRAP?

The easy ways out, if one abandons either dualism or emergentism of the form defined above, are:

1. To cut off dialogue altogether; each cobbler will stick to his or her last and history will show who produces the better shoe.

2. To go for an Anglo-Saxon pragmatic solution which binds all the contradictions together with a dash of common sense and a hearty reef knot — hoping no one will notice it is really a granny. This pragmatic approach is the one that allows, for instance, "causes" to run both ways through the reductive chain: Molecular events can "cause" behavior, and behavioral events or changes in the environment of the organism can "cause" molecular events. Learning "causes" the synthesis of particular proteins which modify synaptic structures such that when these cells are reactivated they "cause" recall. The chain begins and ends in behavior and causes run symmetrically up and down it, presumably passing one another like commuters on parallel escalators.

3. An alternative escape is to classify phenomena as of two types, caused *either* biochemically *or* behaviorally, as in the classic psychiatric distinction between organic and functional psychoses; the two types of phenomena are believed to be able to coexist or even interact, as when diseases are "real" or "psychic" or "psychosomatic." In any event, such "causes" are seen as part of an exhaustively additive category which follows the algebra of biometrics, where variance is always $x\%$ genetic, $y\%$ environmental with $x + y$ approximately equal to 100%. (If there is an interaction between genes and environment, x times y, then this term is always small compared with the others. Not, incidentally, because it has empirically been shown to *be* small, merely because the algebra does not work out if the term gets too big, and nothing in nature should be so complex that simple algebra will not resolve it.)

4. The pragmatists' final escape route is to abandon cause in favor of correlation. The difference between correlations and causes is supposed to be that cause says that "if A, then B follows" in a transitive and time-dependent way. A is a necessary and sufficient precondition for B. A sharpening of this definition would also demand that A be an *exclusive* precondition for B, that is, that it does *not* result in some other phenomenon (B^1) of the same general class as B. By contrast, correlation merely says that when A occurs there is a tendency, other things being equal, for B to occur also. Correlations do not necessarily imply transitive relations (if A, then B; if B, then A); they are probably necessary but need not be sufficient or exclusive (correlations between the rise in the issuing of television licenses and coronary heart diseases in the 1950s in Britain are examples of those that are, one assumes not even necessary but merely fortuitous). So the literature becomes full of correlations between GABA transaminase levels in the olfactory bulb and mouse killing behavior in the rat, dopamine levels in the midbrain and schizophrenia in humans, exposure to training situations and enhanced RNA synthesis in the chick . . . or whatever.

Nothing wrong so far — at least I hope not — as the last could be a precis of the title of several of my own jointly authored papers with one of the editors of this volume! The trouble begins when correlations are seen as sorts of "soft" causes, that when we have observed enough cases in which enhanced excretion of phenylalanine is associated with mental retardation, sickling blood cells with anemia, alcohol ingestion with drunken behavior or cigarette smoking with lung cancer, we will be able to turn the intransitive correlation into a transitive cause; phenylketonuria "causes" mental retardation, hemoglobin substitution "causes" sickle cell anemia, alcohol "causes" drunkenness, and cigarette smoking "causes" cancer, although we may like to evoke a *ceteris paribus* clause and it will not stop some determined people from trying to show that both lung cancer and cigarettes have a covarying prior genetic cause. And, of course, because of the infinite number of correlations possible, there is a corresponding possibility of a steady stream of grant money and output of research publications — even more rewarding than the scope available for systems models — because while the models rarely tell one how to intervene, as they do not deal in putative chemical "causes," the hope is that every so often a correlation will throw up a "real" "cause," and we will be able to correct for phenylketonuria by removing phenylalanine from the diet, for schizophrenia by drugs which act on the dopamine system, and for violence by locking up people with XYY chromosomes.

VIII. FROM CAUSES TO TRANSLATIONS

It will be seen that I am arguing that most conventional attempts to relate biochemical to behavioral events, while they may be very good at generating interesting data or even effective interventions into behavioral processes, do so at the expense of ignoring the conceptual difficulties which underlie the task that they are attempting. In order to integrate the properties of cells and the behavioral patterns into a comprehensive account of the organism, its internal "states of mind" and its externally measurable behavior, the mold in which the efforts at integration are generally cast needs breaking and remaking. I am not sure that I am able to provide a rigorous account of the alternative, but I think I do know where the problem lies and the direction in which its resolution should be sought.

The theoretical confusion lies in the multiple uses of the word "cause" when applied to related phenomena of differing levels of complexity and analysis.[4] Everyone knows what is meant by saying that depriving an animal of food causes it to seek food more actively, or that tasting an aversive substance causes it to avoid that taste subsequently. Analogous to these are such biochemical statements as "the addition of iodoacetate inhibits enzymes of glycolysis, causing an accumulation of glucose" or "the addition of mRNA to an appropriate cell preparation causes the synthesis of protein to commence." These causes are indeed transitive in the way discussed previously and they can be completely contained within a single level of analysis, be this the biochemical or the behavioral. But such causal statements are not of the same type as those which cross levels of analysis: "learning a taste aversion *causes* enhanced protein synthesis in the chick forebrain" or "inhibiting brain protein synthesis *causes* a failure of memory storage," or even "the altered flux of Na+ and K+ ions across the nerve cell membrane *causes* the axon potential." These describe a different type of relationship.

Although the last statements may seem like causal ones, a little thought shows they are not. If "causes" can go "up" a series of levels in a transitive way, they clearly cannot also run "down" the levels in a similar transitive way without paradoxical consequences. Similarly, the temporal element is not of the same form as the first sets of statements; indeed in the last example it is lacking entirely; there is not *first* a passage of ions across the cell membrane and *then* the action potential; rather the events are simultaneous. Further, they are indeed identical; the passage of ions no more *causes* the action potential than the action potential *causes* the passage of ions; there is just one single phenomenon, studied at different levels of discourse. Here we begin to see the clue to the solution to the paradox. Just as "gene" — at least in one of its multiple usages — and "length of DNA" are two names for the same phenomenon (or, indeed, in that hoary old philosophical chestnut, so are the morning and the evening star) and so are "action potential" and "passage of ions" across the cell membrane. Only the language differs: one is physiologese, the other biochemese. So we may also hope to advance toward translations of "memorizing an aversive response" into "modified syntheses of particular proteins in particular ensembles of cells," etc.

It is possible to produce a coherent description of a nervous system in terms of its biochemistry, of regional (statistically described) concentrations of particular molecules, and the dynamics of their interactions. Inputs into this system are the fluxes of molecules from outside, glucose, amino acids, oxygen, hormones; outputs are carbon dioxide, urea, water, other hormones, etc. Regional concentrations of molecules and metabolism also vary

[4] There is another use of the word cause, in logical argument, which is quite distinct from either of these types. I am not concerned to discuss this here.

within the system as a consequence of spatiotemporal events; an influx of acetylcholine into one compartment of the system results in consequent ion fluxes in a slightly different one; substances diffuse or are actively transported from region to region (for instance, down nerve axons, along dendrites) and the rules which regulate such interchanges can be studied rigorously.

A similar physiological model could be produced: The statistical electrophysiological state of the system and its regions could be mapped and changes associated with inputs from sensory nerves, outputs through motor ones, and intraregional synaptic transmission could be described. As with the biochemical description, causal relationships between inputs, changed global and regional states, and outputs could be developed. Note that neither the physiological nor the biochemical description could provide interpretations of the "reason" for particular inputs. Such do not fall within the domain of the open system of the nervous system we are considering here, but could be included by extending the domain of analysis into the environment, although this could still be done within a given level.

Now for the relationship between the biochemical and the physiological description. I argue that between the two there is not a causal, but a mapping, relationship — that is, that there is a correspondence between descriptions of events and processes in the languages of physiology and biochemistry. A complete translation of physiology into biochemistry requires, of course, not merely a correspondence of temporal sequences of events, but also a spatial correspondence — this is why I stressed in the first section of this essay that physiology translated into biochemistry plus morphology. In one sense the mapping that I am describing is isomorphic, although not necessarily one-for-one (as opposed, to one-to-many or many-to-many), and it is no more a *causal* one in its relationship than is the relationship between, say, English and French. The two languages can be translated and are isomorphous, but it is not possible to claim a reductive primacy for one language over the other. The language translation analogy also indicates the limitations imposed on our movement between physiology and biochemistry. While certain types of statements and communications can readily and isomorphically be translated from one into the other language, others cannot. Poetry is an example: it has meaning within any one language which an isomorphous translation destroys, because its meaning is context-dependent — that is, it relies on a broader knowledge of the particular history of a language to convey its full meaning to a listener. So too with the biochemistry/physiology relationship. Certain understandings of the *meaning* of an action potential, say, are only obtainable in physiologese; certain understandings of the *meaning* of conformational changes in membrane proteins associated with ion movements only in biochemese. While anger, love, or learning have their physiological translations, these translations do not have the same context-dependent meanings as in the language from which they have been translated (if they did, it would become possible, if cumbersome, to replace one language entirely by another, and we would merely have cunningly reinstated the old reductionism in a new garb).

There is a further step to take in the argument, however. It is abundantly clear that the domain over which it is possible, in principle, to translate biochemistry into physiology or behavior is limited. Not all biochemistries have physiological translations and not all physiologies have behavioral translations — any more than all physics and chemistries have biochemical translations. Only sufficiently complex chemistry in a sufficiently concentrated spatiotemporal domain constitutes the material for which a biochemical dictionary is possible; and, similarly, only sufficiently complex biochemistries make a behavioral dictionary possible. To speak about the physiology of a stone or even a DNA molecule would be absurd; so too would it be absurd to speak of ameboid psychology. In this critical but strictly limited way, the relationships between the languages of the different levels is asymmetric; all patterns of behavior translate into biochemistry; not all biochemistry translates into behavior. It is this translation, this asymmetry, which enables one to ascribe an order to levels of analysis, to arrange the hierarchy. The higher the order, the richer the meanings and the more compact the description; the lower the order, the wider the domain of relevance within the natural world.

It may be argued that all this is laboring the obvious, but I claim that redefining the theoretical task in the way the preceding paragraphs have attempted avoids a number of the experimental and theoretical minefields sketched out in the earlier sections. It might be asked how it relates to the traditional position taken by Western philosophers of mind (e.g., Vesey 1964); in particular, how does it vary from the "central state materialism" of the identity hypothesis or from epiphenomenalism (Armstrong, 1968). The former states that "mind processes" (the philosopher's term which I have tended to subsume into the term "behavior" in this essay) are "nothing but" or "reduce to" brain processes; the latter that mind processes are caused by brain processes but are irrelevant to the "real" functioning of the brain which proceeds at a neural level, so that brain produces mind like a steam engine produces a whistle, rather than like the kidneys produce urine. The position that I have outlined — I would define it as a dialectical one, though at best it extends dialectical orthodoxy — differs from that of the reductive identity hypothesis because, while claiming that mind and brain processes *are* identical, it insists on the continued legitimacy of the former and resists locating ultimate "cause" in a molecular domain; it differs from epiphenomenalism similarly in that it takes for granted the material reality and causal interrelatedness of the phenomena of mind; no way are they in any sense "epi" to the real, material world.

The tasks of biochemists and ethologists attempting to make a complete description of the organism and its history and relationships then become not those of the search for transhierarchical causes, but for translations between biochemical events and behavioral ones. In such a model the injection of a protein synthesis inhibitor, for instance, does not "cause" an incapacity to store long-term memory; it *does* cause inhibition of protein synthesis, of course, with attendant biochemical sequels, such as changes in ATP levels, glucose and oxygen consumption, amino acid accumulation in the cell, etc. These events may be mapped onto physiological and behavioral ones; at the behavioral level the inhibition is the *same* as the loss of storage ability just as "storage of memory" is the *same* as "such and such a biochemical process" (and the DNA strand is the *same* as a gene).

IX. THEORY INTO PRACTICE

How does this statement of theory affect a practical laboratory program in the search for the biochemical translations of learning? The task of such a program becomes that of defining more precisely, for particular learning tasks, the nature of these biochemical processes and their relationships to the general phenomena of cell biology — that is, what they "mean" to the cell — their time course and their anatomical localization. The assumption is that the biochemical translations of any specific learning process will be similar or identical to those that underlie any other; the specificity of the learning does not translate into specific biochemistries but into specific subsets of cells in which these biochemistries occur; the message is embedded in the connectivity and addresses of the cells and not in unique molecules; for practical purposes, we adopt a Hebbian modifiable synapse model of learning (Hebb, 1949) and assume that the biochemical changes we observe are involved in organizing and directing modifications of connectivity by being part of, or necessary precursors to, changed synaptic efficacy. These changes may take the form of the *de novo* production of synapses, or pre- or postsynaptic architectural or dynamic changes in existing structures, and any learning experience must translate into necessary, specific, and exclusive sets of such changes to adapt Bateson's (1976) formulation. Actually the biochemical methods we adopt are independent of whether the changes are synaptically localized or not; and not all the changes we have studied are, in fact, at the synapse. If the synapses were not involved, our tactics would change, but our strategy would be unimpaired.

There is a further hypothesis that guides our experimentation, and this is that learning, the specific alteration of behavior in an adaptive manner as a consequence of particular experiences, is a special case of the more general plasticity of the nervous system and of behavior. There are many forms that such plasticity can take (q.v., Barlow and Gaze, 1977), and I have postulated (Rose, 1977) that, at the cellular level, neurons may be divided into three general classes: (1) Those that are "hard wired," that is, whose connectivity is epigenetically specified during development and is relatively impervious to environmental change (for instance, the majority of sense-receptor-cortex and cortex-motor-effector connections). (2) Those that are epigenetically plastic, which are programed to respond to particular environments at particular developmental stages by appropriate modulations (for example, the neurons involved in the "ready to go" learning of imprinting. (3) Neurons which are capable of plastic modification throughout life and hence can be involved in adult learning.

It is important to note that the capacity for plasticity is itself epigenetically determined; the dialectic of specificity and plasticity in the development of the organism is genotypically contingent, and transcends the old stereotypes of the nature/nurture dichotomy. But to develop this theme is to move outside the limits of the present essay (see Bateson 1979, 1980). For the present I want merely to return to the more empirical question: is it possible to isolate biochemical events and phenomena which form the translations of the short-term and permanent storage of information of learning and memory? Obviously, I believe that it is, or I would not have spent a large proportion of my researching life attempting to do so. The problems are in part those of experimental design, and in part of methodology. The design questions at the behavioral level revolve around generating experiments that can distinguish, for individuals or groups of animals, between storage processes and the concomitants of storage, such as motor activity, sensory stimulation, and so forth. We do claim that, in our imprinting experiments, we have devised a series of controls that have enabled us to argue, with a reasonable degree of probability, that animals learning the characteristics of an imprinting stimulus show increases in a particular biochemical measure, the incorporation of radioactively labeled precursors into RNA, which seems to be related to the degree of learning the animals show and not to a variety of other behavioral measures such as motor activity, sensory stimulation, attention, arousal, nonspecific hormonal effects, or the mere consequences of light exposure *per se*. It has been possible to localize the site of the RNA changes to a relatively small region of the brain, the medial hyperstriatum ventrale (Horn *et al.*, 1979), and to show that lesions placed in this area after exposure to the imprinting stimulus impair its recall (G. Horn, personal communication).

From the biochemical point of view, the crucial question is the meaning of such changes in terms of cell biology. Changed precursor incorporation could be an artifactual consequence of other biochemical causes, such as enhanced precursor uptake, changed glucose utilization or energy metabolism, or even increased breakdown of preexisting RNA. Criteria can be developed against which to evaluate any proposed biochemical process involved in the storage process (Rose, 1981), but the problem while still partly one of experimental design, is also that of methodology; does one have methods sensitive enough to distinguish between such possibilities when one is dealing with very small differences? If only a tiny number of cells or molecules are involved in a response to the novel stimulation, can one expect to measure such changes against the background noise? To discuss how we have endeavored to make this resolution again takes us too far outside the mainstream of this essay. (The experiments are reviewed in Rose, 1977, 1979.)

In summary, for imprinting and for one-trial passive avoidance learning in the chick, we have been able to identify a series of biochemical events involving the activation of the RNA and protein synthetic systems, synthesis of particular proteins, including the microtubular protein tubulin and synaptic glycoproteins, their export toward the synapses, and the activation of the postsynaptic receptor protein involved in muscarinic cholinergic transmission, in a defined region of the chick brain. We believe that we are on the way to unraveling the biochemical sequence of events that, in temporal terms, matches those of short- and long-term storage, and which provides a mechanism for the modulation of synaptic metabolism, dynamics, and architecture at the biochemical level, which can be translated into synaptic function at the physiological. We also have reason to believe that, in conformity with the predictions advanced earlier in this chapter, the changes that we observe are analogous to those associated with more general expressions of neuronal plasticity. Whether they are the *only*, or even the primary biochemical changes involved, we cannot as yet be certain; we may, however, be reasonably confident that we are working within a framework in which the conceptual apparatus and experimental methodology available to us enables us to approach the task.

Such an ending may seem a modest enough piece of practice to follow such grand theory; perhaps so, and in defense I can only assert that in neurobiology, perhaps more than in any other science, it is both necessary and unconscionably hard to keep moving with one's eyes on the horizon without falling over one's feet.

X. ACKNOWLEDGMENTS

I should like to thank the following for comments on an earlier draft of this chapter, and for helping clarify some of my confusions in attempting to write it; neither the arguments of the paper or those confusions which still remain are of course the responsibility of those mentioned: Martin Barker, Pat Bateson, Bob Burgoyne, Ruth Hubbard, Lou Irwin, Brian Medlin, Ian Morgan, Dan Muir, Lesley Rogers, Hilary Rose, and Fred Toates.

XI. REFERENCES

Armstrong, D. M. (1968). *A Materialist Theory of the Mind*, Routledge Kegan Paul, London.

Barash, D. P. (1976). *Sociobiology and Behavior*, Elsevier, New York.

Barlow, H., and Gaze M. F. (1977), Structural and functional aspects of plasticity in the nervous system. *Phil. Trans. R. Soc. B* **278**:242–436.

Bateson, P. P. G. (1976). Neural consequences of early experience in birds. In: *Perspectives in Experimental Biology, Vol. 1, Zoology*, Spencer Davies, H. (ed.), Pergamon, Oxford, pp. 411–418.

Bateson, P. (1979). How do sensitive periods arise and what are they for? *Anim. Behav.* **27**: 470–486.

Bateson, P. P. G. (1980). Ontogeny. In: *The Oxford Companion to Animal Behavior*. McFarland, D. J. (ed.), Oxford University Press, New York.

Dawkins, R. (1976) *The Selfish Gene*, Oxford University Press, New York.

Engels, F. (1940). *Dialectics of Nature*, Lawrence and Wishart, London.

Feyerabend, P. (1978). *Against Method*, New Left Books, London.

Hebb, D. O. (1949). *The Organization of Behavior*, Wiley, New York.

Heller, A. (1979). *A Theory of Feelings*, Van Gorcum.

Horn, G., McCabe, B. J., and Bateson P. P. G. (1979). Imprinting: an autoradiographic analysis of changes in uracil incorporation into chick brain. *Brain Res.* **168**:361–379.

Irwin, L. N. (1979). Fulfillment and frustration: the confessions of a behavioral biochemist. *Perspect. Biol Med* **21**:476–492.

Koestler, A., and Smythies, J. R. (1969). *Beyond Reductionism*, Hutchinson, Basingstoke, Hants, Eng.

Levy Leblond, J. M. (1976). Ideology of/in contemporary physics. In: *The Radicalisation of Science*, Rose, H., and Rose, S. P. R. (eds.), Macmillan, New York, pp. 136–175.

Monod, J. (1973). *Chance and Necessity*, Cape, London.

Needham, J. (1943). *Time, the Refreshing River*, Allen and Unwin, London.

Oatley, K. (1978). *Perceptions and Representations: The Theoretical Bases of Brain Research and Psychology*, Methuen, London.

Popper, K. R., and Eccles, J. (1977). *The Self and Its Brain*, Springer, Berlin.

Rainbow, T. (1979). Role of RNA and protein synthesis in memory formation. *Neurosci. Res* **4**:297–312.

Rose, S. P. R. (1976). *The Conscious Brain*, Penguin, Harmondsworth, Middlesex, Eng.

Rose, S. P. R. (1977). Early visual experience, learning and neurochemical plasticity in the rat and chick. *Phil. Trans. R. Soc. B (Lond.)* **278**:307–318.

Rose, S. P. R. (1979). Neurochemical correlates of learning in the chick. In: *Neurobiological Bases of Learning and Memory* Tsukada, Y. and Agranoff, B. W. (eds.), Wiley, New York, pp. 179—191.

Rose, S. P. R. (1981). What should a biochemistry of learning memory be about? *Neurosci.* (in press).

Rose, S. P. R., and Longstaff, A. (1980). Neurochemical aspects of learning and memory. In: *Neurobiology of Learning and Memory*, McGaugh, J., and Thompson, R. F. (eds.), Plenum, New York.

Toates, F. (1980). *Animal Behavior — a Systems Approach*, Wiley, New York.

Vesey, G. N. A. (ed.). (1964). *Body and Mind*, Allen and Unwin, London.

Wall, P. D. (1974). "My Foot Hurts Me": Analysis of a Sentence. In: *Essays on The Nervous System*, Bellairs, R., and Gray E. G. (eds.), Clarendon Press, Oxford, pp. 391—406.

Ward, S., Thomson, N., White, J. G., and Brenner S. (1975). Electron microscopical reconstruction of the anterior sensory anatomy of the nematode *Caenorhabditis elegans*. *J. Comp Neurol* **160**:313—338.

Wilson, E. O. (1975). *Sociobiology — The New Synthesis*, Harvard University Press, Cambridge, Mass.

Chapter 9

BEHAVIOR AND THE PHYSICAL WORLD OF AN ANIMAL

Steven Vogel

Department of Zoology
Duke University
Durham, North Carolina 27706

I. ABSTRACT

Much of the behavior of animals constitutes a set of adaptive responses to the physical world in which animals live. The relevant physical parameters are far more numerous than biologists commonly recognize. And the particular parameters relevant in a given situation depend strongly on the size of the organism in question; here our biases as unusually large creatures seriously compromise our facility for any intuitive recognition of just what factors might be important. Parameters such as pressure, temperature, humidity, and fluid motion take on quite different significance, both quantitatively and qualitatively, to organisms of more "ordinary" size than ourselves. But recognition of what might be important to a particular animal requires not an elaborate training in physics, but rather some self-education in how one thinks about the world encountered by the organisms one studies.

II. THE PARAMETERS OF CONCERN

Every animal lives embedded in a multidimensional physical world to which much of its structure and personal activities represent adaptations. This essay is the polemical outburst of a physical biologist impressed with the relative neglect of the physical world by investigators of animal behavior.

In moving beyond descriptive behavioral studies, we commonly go down the reductionist path in search of the elusive "fundamental explanation." The usual (if often inappropriate) test is whether the explanation derives the behavior of the whole from the more deterministic behavior of its parts. Explanations are considered increasingly potent if based on anatomy, physiology, or, best of all, chemistry. And our students are advised that the *sine qua non* of a modern biologist is extensive exposure to the molecular disciplines. The linkage is as potent as it is fashionable, and examples of effective, multilevel studies are numerous.

However, I wish to focus on another dimension of linkage of relevance to the ethologist — linkage involving a different pathway of reduction and considerably further from fashion — the linkage between an animal and the macroscopically relevant physical parameters of its immediate surroundings. Such linkage relates the epiphenomenon of the behavior of some product of natural selection not to increasingly minute components of its compostion and structure, but rather to a host of external physical factors.

Viewed with the special prejudice of a concern with physics, almost all biological structure, at all levels of organization, in some measure reflects physical exigencies. And, unless one takes the preposterous view that behavior and structure are independent, behavior as well must bear a similar relationship with the immediate character of the physical world. That examples of such linkage are less numerous in studies of behavior than in studies of structure is, I believe, an historical accident and is the motivation of the present discussion. Indeed, the traditional concern of behavioral studies with interactions among animals may be just another example of the human interest in events of reciprocal influence, just as we seem to prefer games in which each participant determines his actions by the past and probably ploys of opponents. The less interactive character of animal-environmental behavioral relations has, therefore, less intrinsic fascination. Nevertheless, these relations exist, are fraught with behavioral consequence, and their neglect is surely inadvisable.

The relevant physical factors are many and their attendant applications may be quite subtle. The subleties arise, in practice, from several causes, which I will mention here and examine in greater detail later. First, the very number and interrelated actions of the various parameters confuse matters. Second, the relative relevance of different physical factors is strongly scale-dependent, and our biases are those of quite unusually large animals. Finally, both our senses and our instrumentation seriously constrain our concerns.

It is an interesting exercise to compile a list of physical characteristics of the world around an organism which might be in some way important to it: the extent of even a crude compilation is impressive. Such a list is given in Table I; while I do not claim that it is exhaustive, I do believe that examples

Table I. Bioportentous Parameters

Acoustic transmittance	Heat of combustion	Pressure
Bearing strength	Heat of fusion	Radioactivity
Compressibility	Heat of vaporization	Sound level
Density	Humidity (and dew point)	Sound spectrum
Dielectric constant	Illumination spectrum	Surface tension
Diffusion constant	Index of refraction	Temperature
Electrical resistivity	Luminous flux	Thermal capacity
Electric charge	Magnetic flux	Thermal conductivity
Emissivity	Magnetic susceptibility	Thrml. expnsn. coeff.
Fluid velocity	Modulus of elasticity	Turbidity
Fluorescence	Optical density	Turbulent intensity
Friction of substrate	Osmolarity	Viscosity
Gravitational constant	Photoperiod	Vorticity
Gravitational direction		

exist which illustrate the relevance of each parameter to some animal under some circumstance.

In addition to the values of these parameters *per se*, we should recognize that the spatial and temporal rates of change of many of them constitute biologically relevant variables as well.

Any such list of parameters could (in theory at least) be arranged in descending order of importance in the lives of animals. Similarly, such a list could be arranged in order of the commonness with which the parameters are measured (or considered) in animal experimentation. I believe that it is most unlikely that the two tabulations will be in approximate coincidence. We are biased by our size and our senses, our traditions and our technology. We commonly consider temperature, osmolarity, and turbidity; we usually ignore fluid velocity, modulus of elasticity, and other mechanical properties. We consider the color of an object in the visible portion of the spectrum, but we largely ignore its reflectivity in the ultraviolet (which insects see) or in the infrared (in which half the thermal energy of the sun arrives). But for a proper understanding of adaptations, including behavioral ones, we should not be limited by such essentially accidental biases.

III. THE PHYSICAL WORLD COMES FIRST

It is my own strongly held prejudice that we can derive a certain general guidance from past studies (of the kind which will be recounted here). They argue strongly that one should consider physical factors and the concomitant adaptations of organisms before erecting strictly biological explanations or hypotheses. Does an animal take up a certain posture to control a thermal load or to fool a predator? In general, test the thermal hypothesis first; it will usually be the easier, once an appropriate way of thinking about behavior has been acquired. Do not develop explanations requiring expenditure of metabolic energy until simple physical effects are ruled out. Must a sponge pump all the water it filters or is it designed to take advantage of ambient water currents to increase its feeding efficiency? The latter possibility was long ignored. But, at the very least, this version of Occam's Razor (perhaps we should call it Occam's Scalpel) does presuppose reasonable familiarity with the immediate physical world of the intact organism.

As a first example of behavioral adaptations to the physical world, let us consider a case somewhat removed from the usual concerns of ethologists: the orientation behavior of the leaves of the so-called mimosa tree (*Albizzia julibrissin*), an introduced ornamental common in the southeastern part of the United States. Leaves may adopt any of three distinct postures. In the shade, all of the leaflets of the bipinnately compound leaf are oriented in the plane of the blade and are thus coplanar; the leaf faces any open sky, illumination of the leaflets is maximized, and little light from the sky passes between the leaflets. Exposed to the sun, each leaflet rotates about 90° on its long axis, so, while the leaf as a whole may face the sun, the leaflets lie edge on and the leaf casts almost no shadow; illumination of the leaflets is minimized. At night a third posture is assumed: each leaflet folds flat against the next distal leaflet and all are pressed against the petiole. The petioles are folded against the central rachis, and the entire leaf droops downward, looking somewhat like a horse's tail.

To what might this odd behavior be adapted? A little consideration of radiation and convection easily generates testable hypotheses involving many fewer subsidiary assumptions than would any inquiry into the behavior of the local phytophagous fauna. It is well known that leaves exposed to direct sunlight can get substantially warmer than the air around them. The orientation in sunlight clearly reduces interception of light and almost as certainly increases the efficacy of convective heat dissipation. Leaves exposed to a clear, open sky can get colder than the air around them as a result of radiation from their surfaces. (This, incidentally, is why if the wind is low, frost may form on leaves even when the average air temperature is above freezing.) But, while the sun is a point source of radiation, the open night sky is a distributed sink for radiation; thus the posture which minimizes direct solar exposure would be of little use in reducing exposure to the cold night sky. Instead, by folding against each other, leaflets and petioles minimize their total exposed surface. The basic hypothesis, then, is that orientation behavior reduces extremes of leaf temperature. It has not, as far as I know, been subjected to serious scrutiny; the point here is not whether the hypothesis is correct but rather that it might not be formulated in the first place. Consider an ethologist working with an animal in which such orientation behavior is merely a part of a wider repertoire, an ethologist who had not read Gates's (1962) little book (*Energy Exchange in the Biosphere*) or anything similar. Would this ethologist be likely to relate the orientation behavior to temperature control through adjustments in radiative and convective heat exchange?

IV. SIZE AND THE PHYSICAL WORLD

We are particularly large animals. Whether one considers number of species, number of individuals, or total biomass on earth, humans are in the top fraction of the biggest percentile. The physical parameters of most direct relevance to an organism are very much a matter of how big that organism is. As a result of our large size and the extreme range of sizes over which living organisms extend, our intuitive feeling for physical reality must be considered a most unreliable guide to the world of more "ordinary" sized organisms. An eel and a spermatozoan swim with overtly similar motions, yet the underlying fluid mechanics proves to be drastically different for the two. The "conventional" scheme of the eel would lead to a balance of forces at a velocity of zero for the spermatozoan — it would get nowhere.

Consider, for example, the variation in the effects of *gravity* on terrestrial animals. The gravitational constant does not vary much over the surface of the earth, but its practical effect on an organism is strongly dependent on the size of the organism. Rather roughly, we can distinguish four sorts of animals. The largest kind can sustain injury by a fall from its own height, especially if in motion at the time of the fall — tripping poses perils. Big animals are, in fact, fragile creatures. Calculations based on simple loading suggest that, if a cat needs to devote 13% of its weight to skeleton, a horse should assign no less than 86% of its weight to bones to achieve the same relative strength (the actual figure is 24%), and terrestrial elephants are quite out of the question. Furthermore, the larger animal has farther to fall and, in consequence, arrives at the point of impact with a higher velocity. In short, what is routine behavior for a cat would be quite imprudent for a horse or any conceivable horse-sized mammal.

Not-quite-so-large animals (our cat, for example) can sustain injury by falling, but the height of fall must exceed the animal's own stature. Tripping may be trivial, but the animal must surely look before it leaps. Depending on our age, size, and circumstances, we are at or near the division between these two classes. Went (1968) points out that a two-meter-tall adult man, falling from his own height, will have some 20 to 100 times the kinetic energy to dissipate on impact than will a small child just learning to walk; he calculates that we are about the tallest creature which could be safely bipedal.

For still smaller animals and longer falls, the drag of the animal in motion through the air becomes important. A falling body accelerates less and less as it descends, eventually achieving "terminal velocity" at which the upward force (drag) exactly balances the downward force (weight). There is, however, a general relationship between drag and weight. Drag is roughly proportional to the surface area of an object, while weight is, of course, proportional to mass — and in this case volume. For objects of approximately the same density and shape, the drag-to-weight ratio varies in a manner similar to the more familiar surface-to-volume ration. Since surface area is proportional to the square of a linear dimension and volume to the cube, smaller objects have greater surface-to-volume ratios. Similarly, they have greater drag-to-weight ratios and, as a consequence, lower terminal velocities. A lower terminal velocity means less chance of injury on impact, quite independent of the distance fallen. I once dropped two laboratory mice about 20 m onto a concrete floor. They each adopted a proper parachute-like posture to minimize terminal velocity (had natural selection preceded me in doing the experiment?), were momentarily stunned (and thus recaptured), but were sufficiently undamaged to immediately mate and produce a healthy litter. For animals of mouse-size and smaller, serious injury through falling from any height at all is essentially out of the question. And most terrestrial animals are in this size range — for them falling itself is just not dangerous.

For really tiny organisms, drag forces are relatively monumental. Consider a bacterium (*Escherichia coli*) swimming around at 15 body lengths per second (faster, relatively, than the best human swimmer or runner). Howard Berg (reported by Purcell, 1977) has calculated how far the bacterium would coast if it were suddenly to stop spinning its flagella — it would go about 0.001 nm, or less than a hundred-thousandth of its body length. In its world, as a practical matter, coasting or gliding simply does not exist, so large are drag forces. And thus for very small creatures the direction of fall is not immediately related to gravity. The phenomenon of falling downward has lost its practical significance whether in air or water.

Another physical factor which results in appropriate behavior being a rather size-dependent matter is *diffusion*. Diffusion is some 10,000 times faster in air than it is in water; but even in air, diffusion is a very slow process over distances germane to human perception. (The common demonstration of diffusion in air, releasing perfume in a room, is really a fraud — the spread of the odor is mainly accomplished by convection or bulk flow, not diffusion.) The net rate of diffusion, however, is inversely proportional to the square of the distance from the source, so diffusion may be a powerful agency in the life of a small organism. Taking the extreme case of *E. coli* again, Purcell (1977) has calculated that if the bacterium is living in a uniform medium rather than swimming around to obtain food it is better off just waiting for diffusion to bring the food to it. Locomotion is useful only to search for greener pastures. What cow could wait, between bites, for the grass to grow tall again?

Even the functional significance of environmental *temperature* is quite dependent on the size of an organism. Consider a camel walking across a desert. It could maintain a constant body temperature through evaporative cooling, using precious water in the process; in fact, it does not do so. Instead it tolerates a gradual increase in its body temperature during the day, secure in the knowledge that night will shortly follow, with cooler air and a cold sky for reradiation. The trick works because the camel is large: its surface-to-volume ratio is low, so it warms very slowly (Schmidt-Nielsen *et al.*, 1957). A much smaller diurnal mammal, the desert ground squirrel (Bartholomew and Hudson, 1961) adopts a different pattern of behavior, foraging briefly and returning frequently to its burrow. Why the frequent trips to the burrow? Inside the burrow, the rodent presses its body against the cooler subsurface soil and, through conduction, eliminates the excess heat; its small size dictates the frequency of these cooldown periods. Still further from our intuitive world is the situation of a broad leaf atop a forest tree. It absorbs sunlight and may be $10°$ or more above air temperature. Further temperature rise is limited to some extent by evaporative cooling but to a greater extent by convective air currents (Knoerr and Gay, 1965). I have found that during a lull in the wind the leaf heats (not surprisingly) very rapidly, its temperature rising a degree every few seconds. Thicker leaves appear, in part, to be adaptations to increased tolerance for lulls in the wind — the thicker the leaf, the longer the time constant for temperature rise. The distribution of leaf types among the hollies of North Carolina "makes sense" in terms of local requirements for tolerance of periods of very low wind (Kincaid and Vogel, 1980). Under steady air currents, leaf thickness is thermally inconsequential; but near the surface of the earth, air movement is rarely steady. Might the behavior of small insects reflect the necessity of avoiding extremes of body temperature under circumstances where the sun may shine steadily, but the winds are irregular?

Much of the behavior of an animal involves what might be called *mechanical functions*, and here again the scale of the organism is all important in determining what is possible and appropriate. We are big, bipedal, and manually dextrous, in combination a fine preadaptation to the use of the kinds of tools which pound, chop, scrape, or impale. As Went (1968) has pointed out, all these tools require a massive creature for effective use.

Perhaps we have worried far too much about the mental capacities of other animals in connection with their use or nonuse of tools; perhaps our physical preadaptations are more important, and preeminent among these is our size.

Indeed, it is an easy matter to propose general rules, based on size, for the practical ways one animal might attack another. Kicking and hitting are the prerogatives of large animals, being heavily dependent on kinetic energy; the latter is a product of mass and the square of velocity, so long appendages and a big body to absorb recoil are critical. Similarly, thrown projectiles work mainly for the large: to the requirement for imparting kinetic energy is added the relatively greater drag of a small projectile, further limiting its travel. Perhaps the limitations of a small projectile are best illustrated by considering the so-called *Pilobolus* gun. This fungus shoots (by hydrostatic pressure) a sporangium about 0.1 mm in diameter. (Shooting individual spores would, of course, be completely futile.) Despite an impressive muzzle velocity of 20 m·s^{-1} (45 MPH), the sporangium travels only about 4 m (Buller, 1958). Should we be surprised that aggressive shootings in the world of the small often require direct contact for sufficient momentum transfer and typically employ poisonous projectiles? It is perhaps no accident that in the operation of the nematocysts of coelenterates it is frequently the prey, not the predator, that presses the trigger.

Biting, crushing, and squeezing will work for somewhat smaller as well as for very large organisms. Most mammals, other vertebrates, arthropods, some worms, and some mollusks find these effective, but as such systems become yet smaller the advantage gradually shifts to the defence. The smaller radius of curvature of the prey typically implies a relatively greater strength or resistance to deformation or dismemberment, and it becomes more practical for a predator to burrow into the prey if a mechanical approach is retained. Still, small prey have an inevitable chink in their armor — their surface-to-volume ratio is large, which improves the practicality of a chemical approach. There are innumerable cases in which small predators entice prey into some enclosed place, whether vacuole or stomach, and then pour a fine digestive armamentarium on the mechanically intact prey. We larger predators, dealing with prey of lower surface-to-volume ratio, are more likely to chew food first; the tiny predator cannot, but need not.

Still another physical factor to which the biological response is size-dependent is *surface tension*. In our immediate experience it has little more significance than causing rain to fall in droplets and providing a reason for washing with soap. For smaller organisms surface tension may be a potent part of the physical world. for an ant-sized creature, a bucket would not work — it could not be smoothly filled and emptied since drops of water are just too big. But the bucket may prove unnecessary anyway: the forces of surface tension are great enough relative to those of gravity so drops can with little difficulty be carried around uncontained. Certain insects and spiders can trap bubbles of air underwater; the diving bell above the air need not be airtight but merely have a mesh fine enough so the radius of curvature of a bubble which could pass through would have to be unmanageably small (as will be discussed later). Surface tension thus maintains the integrity of the bubble of air.

Certainly the most impressive behavioral use of surface tension is walking on water, a trick done by a variety of insects (water striders, etc.) and a few other animals. Why does a water strider not fall through? It certainly is not floating, and its density is certainly not much lower than that of water. In practice, surface tension provides an upward force equal to the product of the surface tension of water and the wetted perimeter of the cuticle in contact with the water. If that force exceeds the downward gravitational force, the animal remains astride the surface. It is, by the way, a simple matter to calculate the maximum weight of a person which would permit walking on water. One merely assumes pure water (no detergent pollution) and some foot perimeter; the result is about 10 grams to stand (two feet) or 5 grams to walk (one foot at a time). The theological implications, possibly profound, are beyond the scope of this article.

An interesting way to view a parameter such as surface tension, one of more general applicability than biologists commonly realize, is to concoct a dimensionless index relating the importance of the parameter to another parameter of interest. One can then look at the way in which such an index varies with, say, size, reducing the number of variables one has to think about simultaneously. Taking our example of walking on water, we can examine the ratio between two forces, the upward one due to surface tension and the downward one due to gravity. The former is surface tension γ multiplied by perimeter, and the perimeter (assuming a set of geometrically similar objects) is proportional to length l. The latter is mass times the gravitational constant g or (again for similar objects) density ρ times length cubed times gravity. The ratio (we'll call it Je, for Jesus number) is, then,

$$Je = \gamma l / \rho l^3 g = \gamma / \rho l^2 g$$

This index gives a measure of the practicality of walking on water. Since the surface tension of water, the density of organisms, and the gravitational constant vary very little, we see that Je varies inversely with l^2. Thus for creatures of similar shape, the possibility of walking on water becomes rapidly less as organisms grow larger. Conversely, such behavior represents no startling adaptation for a small animal, and probably only the relative scarcity of smooth water with some decent food source limits its commonness.

The surface tension of water is but one size-related aspect of tension. The latter exists in every stressed membrane from the skin of sharks (Wainwright *et al.*, 1978) to that of cells. A curious relationship exists between the pressure inside shark, cell, sphere, or cylinder and the tension developed in the membrane, a relationship which places certain limits on behavioral possibilities as well as permitting some less than obvious opportunities. The relationship also illustrates another aspect of the dimensional reasoning used earlier — a simple but powerful way of approaching the problems posed by the physical world.

A proper equation describing some relationship in the physical world must be "dimensionally homogeneous," that is, all terms on both sides of the equation must reduce to the same dimensions. Consider the dimensions of tension (force over length) and those of pressure (force over area or force over length squared). Clearly an equation relating tension in a curved wall with pressure within the vessel must have an additional factor with the dimensions of length. For similarly shaped objects all linear dimensions are porportional to each other, so it does not much matter which "length" we use; the radius is convenient for spheres and cylinders. Now the only simple dimensionally homogeneous arrangement for an equation incorporating tension, pressure, and length is the following proportionality:

$$\text{Pressure} \propto \text{Tension/radius}$$

(The constant of proportionality depends only on the shape of the vessel.)

This relationship, then, tells us that a pressure which produces a very small tension in a small vessel will produce a much larger tension in a big vessel. Thus it is harder to start a balloon expanding than to inflate a partially filled balloon further. And, if two balloons are interconnected on arms of a Y-tube, the one which initially begins to expand will expand further as pressure is applied through the open arm of the Y while the uninflated balloon will do nothing. Pressure is more effective as a tension generator in the expanded, larger-radius balloon.

The relationship also tells us that a wall of a given strength, able to withstand a given tension, will be able to withstand a higher internal pressure if its radius of curvature is small rather than large or if it encloses a small vessel instead of a large vessel. In practice, then, a wall of a given thickness will appear stronger when it is the wall of a small vessel. The thin tubes of racing bicycles take far more pressure than those of automobiles; wall tension in an automobile tire, though, is comparable or greater than that in a bicycle tire. An aorta needs a thick wall; an arteriole can withstand similar pressures with a rather thin wall. Small, skinny nematodes can develop within their cuticle enormous pressures (enough to burrow through our muscles), and individual plant cells can withstand osmotically generated pressures of many atmospheres without exploding. In short, one's view of the character and utility of pressurized vessels must not fail to take account of the size of the vessel. And pressurized vessels are ubiquitous — herbaceous stems, sea anemones, worms, the hydraulic leg extension system of spiders, the mantle cavity of squid, and many mammalian penises.

The same relationship explains why small gas bubbles are difficult to form and maintain — the surface tension of water generates enormous pressures inside tiny bubbles, forcing (according to Henry's law) the gas back into solution. Fish find swimbladders of gas space-efficient bouyancy devices but must make appropriate behavioral responses to the metastability of the arrangement. If a fish ascends, the bladder expands, facilitating further ascent; if it descends the bladder compresses, increasing the overall density of the fish and speeding descent. Smaller, mainly planktonic, organisms rarely use gas chambers, most probably on account of their size, but often employ oil droplets to reduce their density. The oil (more dense than gas) is not so space-efficient; but, being much less compressible, it minimizes any problem of metastability.

In emphasizing the importance of size, one cannot easily improve on D'Arcy Thompson's (1942) conclusion to his chapter "On Magnitude," in *On Growth and Form*:

> Life has a range of magnitude narrow indeed compared to that with which physical science deals; but it is wide enough to include three such discrepant conditions as those in which a man, an insect and a bacillus have their being and play their several roles. Man is ruled by gravitation and rests on mother earth. A water-beetle finds the surface of a pool a matter of life and death, a perilous entanglement or an indispensable support. In a third world, where the bacillus lives, gravitation is forgotten, and the viscosity of the liquid, the resistance defined by Stokes's law, the molecular shocks of the Brownian movement, doubtless also the electric charges of the ionized medium, make up the physical environment and have their potent and immediate influence on the organism. The predominant factors are no longer those of our scale; we have come to the edge of a world of which we have no experience, and where all our preconceptions must be recast.

V. BEHAVIOR AND THE FLOW OF FLUIDS

My own major area of interest is the relationship between the structure of organisms and the properties of moving fluids; as with any investigation involving external morphology or structures built by animals, behavioral considerations are frequent intruders. As I shall attempt to illustrate, probably no area of linkage between physics and the world of an animal more easily confounds the intuition of the casual observer. Situations in which the animal is attached or stationary and some wind or current passes over it (as well as the obviously complex cases of locomotion) have been commonly misunderstood, and a lack of physical insight has prevented the recognition of a variety of interesting behavioral and structural adaptations. Perhaps the point can be made most emphatically by a series of examples.

1. The fact that water passes unidirectionally through sponges was first established by Grant in 1825, and this Harvey of the Porifera had appropriately unkind remarks about unsupported statements of Lamarck and Cuvier. By 1864 it was clear to Bowerbank that the motive force propelling this internal current was flagellar rather than muscular, despite the high velocity and volume of water leaving the excurrent opening. Much careful work underlay this conclusion, and its acceptance was not immediate. That flagella can pump water through sponges, even at the usual rate of one body volume every five seconds, is no longer questioned. However, as far as I have been able to determine, an important related question was not asked for another hundred years. If the flagella were inoperative, would water pass through a sponge anyway? Or, to put the matter in more realistic terms, does flagellar action account for *all* of the water passing through a sponge, or can ambient water currents make a contribution to filtration? It appears now (Vogel, 1977, 1978) that not only do ambient currents help, but that the structure of sponges is most exquisitely adapted to take advantage of such currents, with clear functions attaching to a number of previously functionless features. Dynamic pressure on the incurrent openings facing upstream, valves closing incurrent pores lateral and downstream, and suction from the large distal or apical excurrent openings combine to gain advantage from even relatively slow currents. And numerous observations suggest that sponges usually prefer moving water. Why did so much time elapse before someone made a crude model of a sponge, placed it in a current, and watched a stream of dye pass through it?

2. The prairie dogs (*Cynomys ludovicianus*) of the Great Plains of North America are consummate burrowers, even by the high standards of the order of rodents: their typically two-ended burrows in well compacted soil are as much as 3 or 4 m deep and 15 to 20 m long. Studies of their natural history and social organization are not uncommon; they are the subject of at least one entire symposium (Linder and Hillman, 1973) and a thorough investigation by King (1955). Most observers report that entrances to burrows are located in the middle of radically symmetrical mounds and that these mounds generally occur in two distinct geometries, a higher, sharp-edged "crater mound" and a lower, rounded "dome mound." Whether the two forms of mound grace opposite ends of a burrow seems not to have been noted. Mounds are regarded as lookouts or antiflooding devices, and it is generally agreed that following a rain the mounds are

reconstructed, maintaining the dimorphism. Another function of the mounds, however, escaped even casual consideration until recently (Vogel *et al.*, 1973). The oxygen supply in the bottom of a burrow is not unlimited, and diffusion either through burrow or soil can easily be shown to be inadequate for the respiratory needs of an animal in a burrow this deep and long. Wind, though, is a fairly dependable feature of the Great Plains; whenever even a very slight wind blows up, air passes through the burrow system, entering the dome mound and leaving through the crater mound. Model systems in a wind tunnel work nicely, with either the difference in height or shape of the mounds proving adequate to induce a flow through a burrow. The physical basis of the induction is a combination of Bernoulli's principle (the inverse relation between pressure and velocity along a streamline for an ideal fluid) and viscous entrainment (resulting from the resistance of real fluids to high rates of shear). The former, at least, is certainly not a particularly obscure bit of physics.

3. Turret spiders (genus *Geolycosa*) found in eastern North America, particularly on the barrier islands of the Carolinas and in northern Florida, make burrows in open country with sandy soil (Wallace, 1942). A turret spider burrow is a single-ended, almost perfectly vertical tube, 25 to 75 cm deep and 0.5 to 1.6 cm in diameter. The upper portion is lined with silk, and the burrow opens through a small crater of whatever material the spider finds at hand bound together with silk (hence the name "turret" spider). During the day the spider sits at the bottom of the burrow while at night it perches atop the turret feeding on the itinerant fauna. The turret (like the mounds of the prairie-dog) has usually been described as a lookout, although its height is often much less than the spider's own, and it is not clear what use a lookout might serve. My own observations suggest that the spider's legs straddle the turret rather than stand on it.

Consider the following completely physical scenario for the function of the turret (Vogel, in preparation). These spiders, living in very well drained and porous soil, rarely have access to liquid water. Spiders are not as adept at water conservation as the best insects owing, probably, to the different design of their respiratory systems. The air in soil is usually moisture-saturated (even in desert soil a short distance beneath the surface), but *Geolycosa* appear unable to absorb water directly from a saturated atmosphere. During clear days with direct sunlight on the surface of the soil, the temperature in the top few centimeters of soil is usually well above the average air temperature: 55°C is not an uncommon value during the summer. Meanwhile, the temperature near the bottom of a burrow approximates the monthly mean — about 30°C in the summer. Relatively humidity is essentially 100% at the bottom, dropping somewhat near the top. But the dewpoint about 5 cm beneath the top of the turret is typically 12° above the temperature at the bottom, the latter presumably is also the body temperature of the spider. If one dries a spider by about 10% of its body weight and then transfers it back and forth between the climate at the bottom and that near the top of the burrow, a large quantity of liquid water may condense on it, just as water may condense on a container removed from a refrigerator. Placed back in the climate of the bottom of the burrow, the visible water soon disappears, but the spider regains much of the lost weight. Water *may* condense on the spider, but one additional condition must be met. The air at the higher temperature must be moving, even slightly (1 cm·s^{-1}). Otherwise no appreciable condensation appears: diffusion is just too slow, given the rapidity with which a half-gram spider warms up and the low water content of even saturated air. This requirement for moving air, then, gives a function for the turret. Wind across the turret, even gentle breezes, reduces very slightly the pressure within the burrow. Air thus slowly enters the upper portion of the burrow from the sandy soil, providing the requisite movement and probably aiding in maintaining the high dewpoint as well. Again I make the same point about paying attention to the physical world. The climate in the burrow may be determined without especially contemporary instrumentation, and the induction of flow at the turret involves no unusual events. But at least 90 years elapsed between the description of the turret and burrow and the suggestion that the turret, together with a bit of behavior, provides a source of liquid water.

The peculiarly nonintuitive character of the movement of fluids is, as much as anything, an outgrowth of what is perhaps their most characteristic physical property — viscosity. Solids, for all practical purposes, simply do not have this property (or it is so high as to be irrelevant). Instead, they have the related property of elasticity which, in turn, ordinary fluids lack. Elasticity (or, more precisely, elastic modulus) is a measure of the resistance of a material to shearing distortion. Solids resist shear: it takes a force to distort a solid; the greater the force, the greater the distortion. By contrast, fluids, whether gases or liquids, do not care at all how *far* they are distorted — even a trivial force can produce essentially unlimited distortion. But they do care how *fast* they are sheared: to shear a fluid rapidly takes more force than to shear it slowly. The measure of the resistance of real fluids to rate of shear is viscosity. The higher the viscosity of a fluid, the less likely it is to be thrown into eddies and vortices or to be turbulent.

One might expect that the more viscous the fluid, the more force would be required to move an object through it. Oddly enough, this is not always the case. Situations are known in which a higher viscosity slightly reduces drag. A crucial part of the resolution of the paradox was achieved a century ago by Osborn Reynolds, and the scaling parameter he discovered still bears his name. The key to the matter is that the character of flow over an object or through a pipe depends not on viscosity alone but on whether the overall situation is dominated by viscous forces for by inertial forces. The former tend to slow the motion of fluid with respect to solid or other parts of the fluid; the latter tend to keep any element of fluid traveling along its previous trajectory. A dimensionless ratio of inertial forces to viscous forces is thus the single most illuminating index to the character of a flow. Ignoring its derivation, this index, the Reynolds number (analogous to the Jesus number developed earlier) can be written quite simply as

$$Re = \rho l v / \mu$$

where ρ is the density of the fluid, μ its viscosity, v the velocity of the fluid relative to some solid object of interest, and l some characteristic liner dimension of an object immersed in the moving fluid (length, chord, etc.) or the diameter of a pipe through which the fluid flows. Since it is dimensionless, the Reynolds number does not depend on the units in which the variables are expressed as long as the units are consistent (such as all SI).

Let us examine this basic scaling index more closely. At high values of the Reynolds number, inertial forces predominate. Flows are highly "liquid," turbulence is ubiquitous, and drag forces (relative, say, to the mass of an object) are low. At low Re's, by contrast, flows are "sticky": vortices are hard to form and disappear as soon as they are free to do so, turbulence is nonexistent, and drag forces are relatively high. These, then, constitute two rather different worlds, and there is a particularly awkward transition range (Re's roughly between 1 and 10,000) in the middle which just happens to be filled with biological phenomena. Equality of Reynolds number implies that two situations which are geometrically similar will show the same pattern of flow. An object 1 mm in diameter experiencing a flow of 1 cm·s^{-1} in water has a Reynolds number about 10; an object 1 cm in diameter in a flow of 2 m·s^{-1} in air has a Re of about 1000 at ordinary temperatures.

Note also that the Reynolds number contains the ratio of two physical properties of a fluid, density and viscosity. The most immediate consequence is again somewhat counterintuitive. Gases are both less dense and less viscous than liquids, and the gas and liquid of greatest concern to us, air and water, are not exceptional. Thus the ratio of viscosity to density (called the "kinematic viscosity") is not much different for air and water. In fact, the ratio at ordinary temperatures is about fifteen times *higher* for air than for water — air is the more kinematically viscous fluid. So, for an object of given size, exposed to a given velocity, a flow in air will appear 15 times more viscous than a flow in water, with the attendant reduction of turbulence, etc. (For other reasons, drag *per se* will not be greater in air, though.) Or, to obtain a similar pattern of flow, air must be moving, relative to an object, about 15 times faster than the equivalent flow of water.

Curiously and coincidentally, the airflows organisms, commonly encounter are roughly 15 times faster than the flows of water: m·s^{-1} is a substantial water current; 15 m·s $^{-1}$ is quite a stiff breeze.

Finally, the Reynolds number contains the product of length and velocity. In living systems, small size is typically concomitant with a low speed of locomotion or resistance to low-speed flows, whereas larger organisms go faster and tolerate higher currents and winds. Thus the Reynolds numbers germane to small organisms will commonly be very low and those for a large organism will be very high; in practice, size and speed work in concert. As a result, the range of Reynolds numbers relevant to organisms is extreme — at least a dozen orders of magnitude. Still, while the index to the immediate character of the physical world may depend on size, it is not .dependent solely on size but rather (for a given medium) on the product of size and speed or (where media differ) on the Reynolds number as a whole. The same patterns of flow may occur as a result of similarity of Reynolds numbers in such disparate situations as the membrane of a plant cell and an advancing glacier.

At very low Reynolds numbers (below 1) it is entirely reasonable to neglect inertial forces altogether. Under these conditions a number of strange phenomena occur with which we have little perceptual familiarity. Drag, instead of being very roughly proportional to the square of velocity, is instead proportional (almost exactly) to the first power of velocity; instead of being proportional to density, it is proportional to viscosity. Objects (recall Berg's bacterium) come to rest almost immediately after active propulsion ceases. Objects moving with respect to the fluid carry with themselves a large cloud of semiattached fluid which tends to disturb other objects at some distance and to obscure the details of their own surfaces. Swimming "quietly" is impossible — an organism can easily tell when another is moving nearby; and, conversely, a swimming organism can probably often detect nearby surfaces by the distortion of the flow field around it. A small predacious copepod has sensory opportunities of which we have no experience (Strickler and Twombly, 1975). Since surface detail is obscured, drag is not strongly dependent on orientation but rather depends mainly on the total surface exposed to a flow. One cannot use a paddle normal to flow for a power stroke and feather it to a parallel orientation for a recovery stroke. Nor can one make a fast stroke in one direction and recover with a slow stroke in the opposite — with drag proportional to V and not V^2, the total work on each stroke is the same. Without turbulence, mixing is difficult, and stirring a fluid is not necessarily a disordering process — stirring counterclockwise can almost undo the effects of stirring clockwise!

At high Reynolds numbers one might expect to be able to ignore viscous forces. In fact, one cannot; the only simplification is that viscous forces need only to be considered in the immediate vicinity of an object. The problem is that fluid does not "slip" as it passes over a surface: the fluid just adjacent to the surface is stationary and a velocity gradient exists in the fluid between the surface and the so-called "undisturbed fluid" some distance away. Within the region of the gradient (the "boundary layer") shear rates are high as "layers" of fluid are in effect sliding across one another, and so viscous forces are significant near surfaces at any Reynolds number. The no-slip condition and the existence of boundary layers are fraught with biological and behavioral implications. An organism can gain protection from wind and currents by remaining close to a surface. Conversely, to feed or shed gametes into a current, an organism must protrude at least partway through the local boundary layer. Thicknesses of boundary layers can be (roughly) calculated and vary inversely with the square root of the Reynolds number.

Perhaps the most commonly considered consequence of a fluid moving across an object is the drag experienced by the object. An attached organism must mechanically resist this force and perhaps be deflected by it or reorient as a response to it. Properly designed and maintained irrigation canals afford no place with sufficiently low drag on snails for attachment to be maintained and can thereby break the infective chain of schistosomiasis (Jobin and Ippen, 1964). Drag, however, depends in a complex (and again counterintuitive) way on shape, orientation, and Reynolds number. At very low Reynolds numbers drag is mainly a matter of how much surface an object exposes — "skin friction" is the predominant component. A sphere has about the lowest drag per unit volume contained of any shape; and, not surprisingly, many small locomoting organisms are nearly spherical.

By contrast, at intermediate or high Reynolds numbers, shape and orientation matter a great deal, and a sphere is far from a good low-drag shape. Another sort of drag, "pressure drag," has become important. The latter results from the fact that fluid moving around, for example, a sphere, does not move symmetrically from front-center to rear-center. In actuality, it detaches or "separates" near the widest part, leaving a region of irregular eddies just behind the sphere. Thus the work put in to accelerate the fluid out and around the sphere is not regained as the fluid decelerates again behind and is instead dissipated in the wake. Smooth deceleration and recapture of the work may be accomplished by using, not a sphere, but a so-called "streamlined" shape: rounded (nearly hemispherical) in the front, but gently tapering to a point behind. At high Reynolds numbers a streamlined shape may have only 1% of the total drag of a sphere of the same cross-sectional area. Thus streamlining is universal among large, swift fliers and swimmers and is clearly a good thing. But at low Reynolds numbers it may, by exposing additional surface, actually make matters worse.

A final and particularly elegant example of a situation in which fluid mechanics comes to the rescue of natural history and behavior is provided by O'Neill's (1978) study of feeding in the sand dollar *Dendraster excentricus*. This particular sand dollar stands upright with the anterior edge of the test supported by sand rather than lying flat on its oral surface as do other sand dollars. The animals are highly gregarious: within beds individuals may (without touching each other!) reach densities of 1000 per m^2, each oriented parallel to the water flow. Previous explanations of the upright posture and dense aggregations were unsatisfactory; O'Neill proposed and demonstrated that these cambered discs constituted lifting bodies (lift in the aerodynamic sense, here directed laterally rather than upward). She showed that their feeding efficiency was increased by the effects of neighboring animals in directing currents toward the oral surface; each animal takes advantage of the curvature of the streamlines induced by its neighbors.

VI. REMARKS IN CONCLUSION

At one level, this discussion does no more than belabor the obvious — organisms live in a physical world and much of their structure and behavior represents adaptations to it; we may properly call it "linkage" in much the manner that we can recognize linkage between behavior and structure, behavior and neurophysiology, behavior and biochemistry. The more important point is the polemical one: the linkage is there and we are likely to be led seriously astray if our ethology does not recognize it. So much for diagnosis. Prescription is not quite so simple. Different people look at the world through different filters, and thus I find the company of ethologists (outdoors at least) endlessly fascinating since their filters are different from my own. A decent appreciation of the physical world is largely a matter of modifying one's outlook. There are a few helpful sources. I have already mentioned Gates' book on energy exchange through physical processes. On the flow of fluids, Shapiro's (1961) little book, *Shape and Flow*, is both engaging and informative; on the behavior of solid materials, Gordon's *Structures* (1978) is a joy to read. A few other such works may exist although unfortunately the genre is far too rare.

Beyond reading, it is very much a matter of practice. With my interest in the flow of fluids, I sit on the edge of a stream and see a certain world. Submerged vegetation waves and indicates the streamlines of the invisible current, and floating leaves ripple on the surface, with dragonflies patrolling above. Butterflies make their upwind approach to flowers, showing me wind direction at some distance, while hoverflies are fixed as if impaled by a shaft of sunlight, giving only a twitch with each shift of the breeze — a rough and ready differentiation of what the butterflies say. Whirligigs and waterstriders dimple the surface, demonstrating that surface tension is normal and oil and detergent reasonably rare. I can, with a little imagination envision the warm convective plumes around the heated leaves of sun-lit trees and bushes, the cool air moving with surface water of the stream. And I can review the unseen motions, the filtration by the silken webs of trichopteran larvae, the movement of fluids in the vessels of plants, the cyclosis in cells, my own convective plumes, respiratory airstream and circulation, the peristalsis, micturition, and flatulence of recent meals. With more effort I can focus on the radiative and conductive energy exchanges; it takes still more effort for less rewarding results for me to contemplate sound and the acoustic environment. Again, we each have our biases, predilections, and expertise. The selective process, though, deals with all parameters it finds relevant to survival, and we should make our best efforts to do likewise.

VII. REFERENCES

Bartholomew, G. A., and Hudson, J. W. (1961). Desert ground squirrels. *Sci. Am.* **205**(5): 107–116.

Bowerbank, J. S. (1864). *A Monograph of the British Spongiadae,* Vol. 1, The Ray Society, London, 290 pp.

Buller, A. H. R. (1958). *Researches on Fungi,* Vol. VI, Hafner Publishing Company, New York.

Gates, D. M. (1962). *Energy Exchange in the Biosphere,* Harper and Row, New York, 151 pp.

Gordon, J. E. (1978). *Structures, or Why Things Don't Fall Down,* Plenum Press, New York, 395 pp.

Grant, R. E. (1825). Observations on the structure and function of the sponge. *Edinburgh Phil. J.* **13**:94–107; 333–346.

Jobin, W. R., and Ippen, A. T. (1964). Ecological design of irrigation canals for snail control. *Science* **145**:1324–1326.

Kincaid, D., and Vogel, S. (1980). Variations in form and thermal properties of *Ilex* leaves from various habitats in North Carolina. (Submitted for publication.)

King, J. A. (1955). Social behavior, social organization, and population dynamics in a black-tailed prairie-dog town in the Black Hills of South Dakota. *Contr. Lab. Vertebr. Biol. Univ. Mich.* **67**:1–123.

Knoerr, K. R., and Gay, L. W. (1965). Tree leaf energy balance. *Ecology* **46**:17–24.

Linder, R. L., and Hillman, C. N. (1973). Proceedings of the Black-Footed Ferret and Prairie-Dog Workshop. South Dakota State University, Brookings, 208 pp.

O'Neill, P. L. (1978). Hydrodynamic analysis of feeding in sand dollars. *Oecologia* **34**:157–174.

Purcell, E. M. (1977). Life at low Reynolds number. *Am. J. Phys.* **45**:3–11.

Schmidt-Nielsen, K., Jarnum, S. A., and Houpt, T. R. (1957). Body temperature of the camel and its relation to water economy. *Am J. Physiol.* **188**:103–112.

Shapiro, A. H. (1961). *Shape and Flow,* Doubleday Anchor Books, Garden City, New York, 186 pp.

Strickler, J. R., and Twombly, S. (1975). Reynolds number, diapause, and predatory cope-
    pods. *Verh. Internat. Verein. Linmol.* **19**:2943—2950.
Thompson, D'Arcy W. (1942). *On Growth and Form*, Cambridge University Press, Cambridge,
    1116 pp.
Vogel, S. (1977). Current-induced flow through sponges *in situ*. *Proc. Nat. Acad. Sci.* **74**:2069—
    2071.
Vogel, S. (1978). Evidence for one-way valves in the water flow system of sponges. *J. Exp.
    Biol.* **76**:137—148.
Vogel, S., Ellington, C. P., Jr., and Kilgore, D. C., Jr. (1973). Wind-induced ventilation of the
    burrow of the prairie-dog, *Cynomys ludovicianus. J. Comp. Physiol.* **84**:1—14.
Wainwright, S. A., Vosburgh, F., and Hebrank, J. H. (1978). Shark skin: function in loco-
    motion. *Science* **202**:747—749.
Wallace, H. K. (1942). A revision of the burrowing spiders of the genus *Geolycosa* (Araneae,
    Lycosidae). *Am. Midl. Naturalist* **27**:1—62.
Went, F. W. (1968). The size of man. *Am. Sci.* **56**:400—413.

Chapter 10

ESCALATED FIGHTING AND THE WAR OF NERVES: GAMES THEORY AND ANIMAL COMBAT

Peter G. Caryl

Department of Psychology
University of Edinburgh
Edinburgh, Scotland, U.K.

I. ABSTRACT

In the last decade, games theory has been applied to the analysis of animal combat — a topic that has been of interest to ethologists for many years; but the process of bridging the gap between these two approaches has only just begun. This chapter complements an earlier paper in which I asked whether games theory could throw new light on some classic ethological data on communication by agonistic displays. In the present chapter, I use the problems raised in ethological descriptions of escalation in animal combat to explore general questions about the nature of explanation in games theory, and to analyze some features of the relationship between two major categories of theoretical model. I argue that the models which appear at first sight to be most directly useful to an ethologist interested in escalation are not, in fact, relevant to this topic, but that a model recently developed by Bishop and Cannings may prove an appropriate model of the phenomenon.

II. INTRODUCTION

"Why do animals prefer pacifism and bluff to escalated fighting?" (Wilson, 1975, p. 247).

A full answer to Wilson's question must contain two components: (1) an explanation of the mechanisms by which animals settle contests using threat or ritualized combat; and (2) a statement of the way in which these mechanisms evolved, a statement which must specify their selective advantage to the individuals concerned.

Ethologists have often argued that the ritualized behavior that is evident in many species in the settlement of disputes is important because it allows communication of differences (e.g., in motivation or strength) between rivals, and that it is this sharing of information which allows the dispute to be settled without violence. However, although ethologists have investigated the mechanisms involved in some depth, the selective advantages have rarely or never been spelled out in detail [except in frankly group-selectionist terms—see, e.g., Harrison Matthews (1964) and particularly the first comment in the discussion of his paper].

While ethologists have tried to frame hypotheses which will explain (*post hoc*) the way in which animals do behave in contests, games theorists have tried to work out rules governing the way in which animals ought to behave in these encounters, when the process of natural selection and the separate interests of the rivals are considered explicitly. One of the most obvious difficulties in breaking down the barriers between ethology and this new approach is the extreme simplicity of the theoretical models; these carry an air of unreality to the ethologist, aware of the complexities of real behavior. Despite this, games theory is of great potential value to ethology. Specification of the selective advantage of particular tactics (e.g., the adoption of relatively aggressive or pacific roles in an interaction) is not easy, since it will depend on what tactics are adopted by other members of the population. In the early 1970s, Maynard Smith was able to show that this difficulty can be overcome by the application of games theory (Maynard Smith, 1974; Maynard Smith and Price, 1973). Dawkins and Krebs (1978) have recently provided a general introduction to this approach, and have argued that it is widely applicable in the analysis of communication.

In an earlier paper I took a prediction from one of the simplest games-theoretic models, the War of Attrition, and showed how it could stimulate the analysis of data at a behavioral level (Caryl, 1979). In this paper I shall look at some theoretical issues from an ethological point of view. One problem which faces those who attempt to integrate the two approaches is the diversity of theoretical models, and the absence of discussion by theorists of the relationships between different models, the contexts in which particular models might be applied, or of the relative importance of the assumptions on which they are based and the robustness of their conclusions when these assumptions are not completely met. It is here that ethologists may make an important contribution to the development of this area of games theory. The significant features of a model often appear very much more clearly when an attempt is made to apply it to real behavior.

In this chapter I shall focus on the topic of escalation in agonistic encounters, to illustrate the way in which consideration of a real ethological problem can help to set theoretical issues in perspective. Examination of the relationships between two categories of theoretical models — one originally designed to deal with display alone, the other to account for the coexistence of "conventional" and "escalated" combat — leads me to ask whether the ethological evidence can allow us to choose between them and suggest directions for theoretical development.

III. ESCALATION IN CONTESTS

It is clear that the question of why there are so many aggressive displays and why an aggressive interaction may progress through several different levels of intensity are questions that games theorists must answer if their models are to remain of interest to ethologists.

Where a species uses several different aggressive displays, ethologists have sometimes argued that these indicate different levels of escalation of the combat. In fighting over nest-burrows, female iguanas (*Iguana iguana*) use six distinct displays, and also bite; victory is sometimes achieved by display alone, but the decision may not be reached until one animal has bitten its opponent. Rand and Rand (1976) used the temporal patterning of the displays, and a subjective assessment of the energy expended in each display, to rank them from Mouth Open to the most escalated, Lunge-Huff (a rapid approach toward the opponent, from a short distance away, with open mouth and a loud "Huff" call). They argued that while the most escalated aggressive acts (Lunge-Huff and Bite) were more likely to cause the opponent to retreat than the weaker displays, the latter were sometimes effective, and an individual had to balance their lower chance of success against the lower energy they required in deciding which display to use.

Games theorists have used the term escalation in a rather different way from ethologists, usually in the form of a simple dichotomy between "escalated combat" and "conventional combat" (or "display"). Although this simple classification can be applied to some animal contests it misses the progressive quality of the changes in interaction that many species show in the course of a fight. For instance, in the Fighting Fish (*Betta splendens*), the interaction can be divided into two phases: threat, and biting combat proper; but the first exchanges of bites are brief, and Simpson (1968) has emphasized that the quality of the interaction changes little at this point. (Later on, if the encounter is very prolonged, display may degenerate.)

The distinction between the game theorists' and the ethologists' basic conceptual models is well illustrated by Laudien's schematized picture of

fighting fish combat (Fig. 1: Laudien, 1965). While the interaction is divided
into phases of threat and biting combat, there are progressive changes in the
behavior within each phase, and Laudien has used the typical order of appearance of behavior patterns to link them to particular levels of "aggressiveness," symbolized by the height of the columns, which is assumed to increase progressively during the fight.

This concept of a series of behavior patterns revealing gradually increasing levels of aggressiveness is of course an idealization. The suggestion
of qualitative differences in the agressiveness of animals showing different
behavior patterns (e.g., Gill Cover Erection and Tail Beating) cannot really
be supported by the data. In fact, it may be difficult to associate a particular
behavior pattern with even a single tendency (a term here used to represent a
drive-like entity which must, however, be measured in an objective way), let
alone with a particular intensity of this tendency (Simpson, 1968). But even
so, the progressive changes in the duration or frequency of behavior patterns in the interaction are loosely compatible with the idea that the comba-
tants become progressively more aggressive until one gives up.

How is this decision about winner and loser reached? What part does
escalation to the level of biting combat play in the decision?

Not all encounters reach the stage of biting. In species such as *Tilapia
mossambica* (Neil, 1964) it occurs only when the opponents are closely
matched in size, and such evidence suggests that this stage is reached when
the encounter has not been (or cannot be) decided as a result of interaction
during the preceding phase of threat. In some species, a conspicuous feature

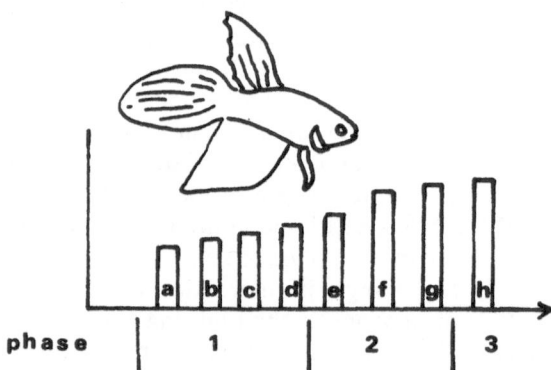

Fig. 1. Escalation in Fighting Fish combat (after Laudien, 1965). The fight is divided into
three phases, (1) threat, (2) close-range combat; (3) end-phase. Activities are (a) stops with
medial fin erection, (b) broadside swimming, (c) pectoral fin-flickering, (d) gill cover erection,
(e) tail beating, (f) ramming, (g) mouth-fighting, (h) chasing/fleeing. Ordinate is the estimated
"aggressiveness" of each type of behavior; abscissa is the time since start of the fight.

of the fight is the way in which an individual matches its level of escalation to that of its opponent; initial discrepancies are reduced by a feedback process [Dow *et al.*,1976, on the killifish (*Aphyosemion striatum*)], and for much of the encounter the behavior of the loser-to-be is not grossly different from that of the winner-to-be, although differences may be detected in subtle measures of the behavior (Simpson, 1968, for *Betta*). In Simpson's work, the eventual loser was as likely as the eventual winner to be the first to bite. In the fighting fish, biting is not (as in the Rands' iguanas) a move that will guarantee victory.

Ethological work on species such as the *Betta* and *Aphyosemion*, where opponents match their level of escalation to that of their rival, has suggested that the matching process allows communication between individuals, and is an important factor in deciding the outcome of the fight. Simpson likened the two opponents to two bidders at an auction each trying to match or outbid his opponent. The question of whether this process is communication in the strict sense of the term will be discussed at the end of the chapter.

IV. GAMES THEORY AND ANIMAL CONTESTS

A. Some Basic Ideas

In its application to animal behavior, games theory assumes that behavior has costs and benefits which can be quantified in units based on the contribution to the individual's reproductive fitness. For instance, the victor in a dispute might acquire a food item, a potential territory, or a mate, and in each case the result is assumed to produce a benefit, V, which can be quantified in units of reproductive fitness. The disadvantageous consequences of fighting (ranging from death or serious injury to exhaustion or mere waste of time) are subsumed under the general term "costs," represented variously by D, T, or as a function of V in this chapter, again quantified in units of fitness. One of the most useful discussions of the reasons for the choice of this unit as a "common currency" is given by McCleery (1978, pp. 380-391).

Each model also includes assumptions about tactics that an individual is allowed to adopt in a dispute, and about the chances of victory or of incurring costs while using these tactics. The benefits that accrue to an individual adopting particular tactics will depend on what tactics are adopted by other members of the population, and for some models, this frequency dependence leads to perpetual change in the proportion of individuals adopting particular tactics. But Maynard Smith has shown that for certain models of aggressive interaction, there exist *Strategies* (combinations of tactics — gambits might be a more appropriate term — in specified proportions) which are *Evolutionarily stable* in the sense that once they have arisen in a population, they cannot be displaced by a mutant using the same gambits in different proportions (Maynard Smith, 1974, 1976; Maynard Smith and Price, 1973; Maynard Smith and Parker, 1976). [If the theorist changes the rules of the game by allowing the mutant to adopt different tactics, it may be able to displace the members of an evolutionarily stable strategy (ESS) population. The statement that a particular strategy is an ESS is only relevant within the specific universe of possibilities considered by the particular model.]

To return to Wilson's question, "Why do animals prefer pacifism and bluff to escalated fighting?," Maynard Smith and Price (1973) showed that for a species capable of both ritualized and escalated fighting — the latter likely to lead to serious injury to one opponent — one ESS was to adopt the ritualized level, only escalating if the opponent did so first.

From the assumptions of the model, we can calculate the *Utility* (the balance of benefits and costs) of playing a particular strategy against another.

If $E[Iv(J)]$ is the utility of playing tactic I in a population playing J, the rules defining that I is ESS are as follows:

Either animals playing the ESS do better against others playing the same strategy than an animal adopting any other strategy can do against one adopting the ESS:

(A) $E[Iv(I)] \geqq E[Jv(I)]$

or if the previous utilities are equal, as they will be in a mixed ESS (see below), the return to an animal playing ESS against a mutant strategy must be greater than the return to an animal playing the mutant strategy against another which also adopts the mutant strategy

(B) $E[Iv(J)] > E[Jv(J)]$

Theorists have developed models to account for both symmetrical and asymmetrical contests. In the former, the opponents are equally matched, and equivalent in every other way; in the latter there is some asymmetry, either related to their ability to win the fight, or else some other difference, such as first arrival and second arrival at a food source, which distinguishes them. In this chapter I shall deal only with symmetric contests; these were the first to receive detailed theoretical analysis, and have provided a basis from which the later analysis of asymmetric contests (e.g., Maynard Smith and Parker, 1976) has been developed.

B. A Simple Model: The War of Attrition

The War of Attrition is one of the simplest models that have been considered. It represents a contest which is settled by display alone. In the model, individuals are imagined to show their threat display (Fig. 2) at constant intensity until one gives up, leaving the other, which was prepared to go on at this point, as the winner. In this game Maynard Smith (1974) showed that the ESS is to choose the duration of the display, X, according to the negative exponential distribution

$$P(X) = (1 / V) \exp(- X / V) \tag{1}$$

shown in Fig. 2.

A result which is important for subsequent arguments is that the average cost of the contest (owing to the time wasted in the display) under this model is equal to $V/2$, where V is the gain from victory (Maynard Smith, 1974).

In his analysis, Maynard Smith tacitly assumed that the cost of display would be a linear function of its duration. But this assumption may not be valid, even where the display is of constant intensity. To display, an animal must neglect other activities, and the cost per unit time of neglecting feeding, for instance, would be expected to rise sharply as the animal's food deficit becomes larger. In another context, there is evidence for "cost functions" which are quadratic (review by McCleery, 1978).

Fig. 2. The War of Attrition. The graph inset shows the distribution of contest lengths at the ESS: $P(X) = (1/V) \exp(-X/V)$, where X is the duration of the contest and V is the prize.

The assumption that the cost of display is linearly related to its duration is not necessary. Norman *et al.* (1977) and Bishop and Cannings (1978) have shown that the model can be extended very easily to deal with cases in which the cost per unit time changes during the course of the display — equation (1) above is rephrased in terms of a "cost function" that relates the cost of the display to its duration.

The War of Attrition is an example of an important class of models that I shall call "continuous models"; the cost of a contest, dependent on its duration, is continuously variable. An alternative type of model is what I shall call the "discrete model," and is exemplified by the model used by Maynard Smith and Price (1973). (In this model, the contest could be fought at two distinct levels of escalation, and the most important factor in bringing it to an end was serious injury, which produced a large, discrete, increment in the cost of the contest and caused the injured animal to cease fighting.)

Games theorists have used discrete models to model escalation in animal contests, assuming that the escalation involves a series of steplike changes in the intensity and potential danger of the interactions. However, in my discussion of the ethological evidence, I emphasized the progressive nature of the escalation in many animal contests, a quality which suggests that a continuous model might be more appropriate. The effects of changes in the vigor or intensity of display, and perhaps also changes in the risk of serious injury, might in principle be accounted for in terms of an appropriate cost function.

But continuous models depend critically on assumptions about the function relating the costs to the duration of the contest. As yet, we have no information about the shape of these cost-functions, although we can make guesses from the distribution of contest lengths.

Since the models can in theory predict almost any distribution of contest lengths, and in practice cannot be provided with realistic estimates of the cost function, are they not too powerful? Why try to represent variation in cost (due for example to escalation in the intensity of the combat) by a hypothetical cost function when perfectly good models exist which were designed to deal with differences in the level of escalation of the contest? Why not use discrete models?

In the next part of this chapter I shall look critically at some discrete models, and at the problem of defining their predictions on the question of escalation. My analysis leads me to suggest that while continuous models produce mixed ESSs which may be unstable in some contexts (see Riley, 1979) they offer many advantages as a basis for theoretical development in the attempt to explain the patterns of escalation that have been of interest to ethologists.

C. Models with Escalation

The occurrence of escalation of the sort discussed in Section III raises several questions:

1. What are the functional differences between combat at different levels of escalation?
2. What factors determine the rules an individual should adopt in deciding whether and when to move from one level of escalation to another?
3. In what way might the system allow communication?

Initially, we shall be concerned with the first question only.

It has been widely accepted that one functional difference between escalated combat and threat is the cost of the behavior. Escalated combat involving physical fighting is often clearly more energetic than the displays that may begin the contest; the individuals may also be less responsive to events outside of the combat, and hence more susceptible to predation; and the combat may produce injuries which carry a cost in terms of a period in which the efficiency of some or all types of behavior is decreased, even if they do not result in death. Recent work has corroborated the earlier analysis of Geist (1971) in showing that serious injury and death do occur as consequences of fighting, and provides some support for the idea that the risk from escalated combat may be considerable.

In some models (e.g., the Hawks and Doves model described below) it is argued that this fact is sufficient to explain the persistence of some form of low-cost nonescalated combat within the population: although an animal which is prepared to escalate further than its rival may win the contest, the escalator will pay a heavy cost when it meets another individual like itself, and as this is likely in a population composed mainly of escalators, pacific individuals will be at an advantage. For the moment, I shall not discuss whether the probability and severity of the injuries recorded in the data of Geist and other workers are compatible in detail with the assumptions of these models. These data are dealt with in Section E below.

It should be noted at this point that Geist (1974) has been highly critical of the approach of games theorists, and has developed the idea that ritualized combat involves an interplay of defensive and offensive behavior, the need for the former holding the latter in check and being responsible for the "formalized" nature of animal combat (Geist, 1971, and see especially Geist, 1974). Although Geist (1974) was correct to point out that the theorists' early comments (about the reluctance to strike "foul blows" in animal combat, for example) were based on outdated ethological literature, it is not clear that this objection is sufficient grounds to dismiss games theory

— in fact, the choice of a balance between risky offensive acts and prudent but unproductive defensive behavior during the contest seems exactly the sort of problem that should be modeled in games-theoretic terms.

D. Hawks and Doves

The simplest discrete model is the celebrated Hawks and Doves game (Maynard Smith, 1976; Maynard Smith and Parker, 1976); the payoff matrix is shown in Fig. 3.

Hawks always escalate, and win over Doves, which refuse to do so. But when two Hawks meet, they fight until one is seriously injured, while when two Doves meet, they settle the combat randomly with no cost. (In this form of the model, it is assumed that the contests between Doves are so brief that time wasted in these contests does not impose a cost in these encounters.)

If the cost of injury D is so great that it exceeds the value of the prize, V, then Hawks cannot exclude Doves from the population: the ESS is a mixed equilibrium with

$$p = V/D$$

where p is the proportion of Hawks. If $D \leqq V$ then all animals are Hawks.

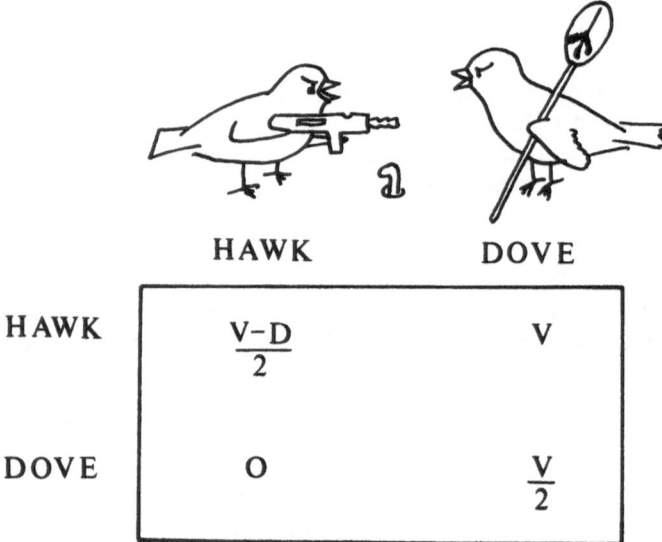

	HAWK	DOVE
HAWK	$\dfrac{V-D}{2}$	V
DOVE	O	$\dfrac{V}{2}$

Fig. 3. The Hawks and Doves game. The payoffs in the table are to the tactic in the row when played against the column. V is the gain from victory, D the cost of injury. For the matrix shown, which ignores the cost of threat, the ESS is to escalate with probability $P = V/D$ (when $V < D$). If threat imposes a cost T on each opponent, the probability of escalation becomes $P = (V + 2T)/(D + 2T)$.

Can a model of this sort, perhaps with a more complex arrangement of levels of physical escalation and the costs associated with them, provide an explanation for the occurrence of different levels of escalation in contests between real animals?

I think the answer to the question is "No," and that once what I shall call the decision rules of such models are made fully explicit, it is very difficult to decide what the biological implications of their results are.

E. "Explanation" by Models

At this stage I shall digress briefly to look at the more general question of whether games-theoretic models explain behavior, and if so where their explanatory power resides. This is a point that has caused some controversy; for instance, Oberdieck and Johnson (1978), discussing some of Maynard Smith's (1977) models of parental investment, have argued strongly that his models have no explanatory power, and by implication criticized this general approach to the analysis of behavior. I believe that although their criticism is unjustified in this particular case, the nature of the "explanation" provided by games-theoretic models deserves much more discussion than it has so far received.

The degree of explanation provided by a mathematical model depends on the balance between the complexity of the mathematical manipulation in the model and the complexity and number of assumptions that it contains.

At one extreme we have models, such as Maynard Smith's models of parental investment, which involve almost no mathematical manipulation. Such models make explicit the relationship between the behavior they predict and the underlying assumptions (in the sense that the conspicuous mathematical features of the model are all direct consequences of the initial assumptions), rather than explaining the behavior. It is Oberdieck and Johnson's failure to appreciate this aim, explicitly stated by Maynard Smith (1977, p. 1), which must lie at the root of their criticisms of his models.

At the other extreme, we have models of considerable mathematical complexity, which have results which are not obvious from the initial assumptions. Because of this, there is a sense in which the models might be said to explain the behavior. Many of the models of aggressive behavior are of this type.

The reader should note that the distinction made here is not the same as that drawn by previous authors, e.g., Dawkins and Dawkins (1974) in discussions of the explanatory value of models in ethology. Dawkins and Dawkins discuss models which yield quantitative predictions which can be compared in detail to experimental results, and point out, for example, that even if a model can predict the experimental data, it may have little explanatory value if it has an excessive number of parameters, or is very sensitive to the choice of parameter values. Similarly, they argue that we should look for models giving strong predictions (predictions which are easily falsified, for instance because they are quantitative and detailed, or else because they are inherently unlikely), rather than those that give only weak predictions. In the present discussion, the distinction that I have made is closer to that between trivial and nontrivial results in mathematics. Game-theoretic models of aggressive behavior have not yet produced quantitative predictions about aggressive behavior which can be applied to the available data.

In models of aggressive behavior, the most important assumptions are the decision rules which allow us to determine what the outcome of a contest would be. Thus, when Hawk meets Hawk, we specify:

1. They shall fight until one is injured (selected randomly if equally matched).
2. The injured one stops fighting at once.
3. The injured one pays a cost D. The victor gets V, the prize.

The payoff matrix, on which the mathematical manipulations are based, is a direct consequence of these rules, and the value of the results depends on the biological relevance. Problems arise in interpreting the implications of discrete models such as Hawks and Doves because the decision rules are not completely stated, and the biological implications of these rules are not spelled out.

F. A Benefit of Destructive Combat?

I want to illustrate this point by considering an argument used by Treisman (1977). Figure 4 shows the average return to an individual in a Hawks/Doves population as the value of D changes relative to V. For a mixed population, this curve would apply to either Hawks or Doves — the return to each type would be equal in the ESS.

In his paper, Treisman was primarily concerned with investigating what would happen if he introduced one or two new types of mutant individuals into a Hawks/Doves population. He considered a new type that fought even more dangerously than existing Hawks, and won over them without damage, and showed that rather than a complex balance between the Doves, the existing Hawks, and these new "Super-Hawks," the population would be a simple mixture of Doves and the new mutant. Hawks, if they were present, would be forced to display only, and thus indistinguishable from Doves.

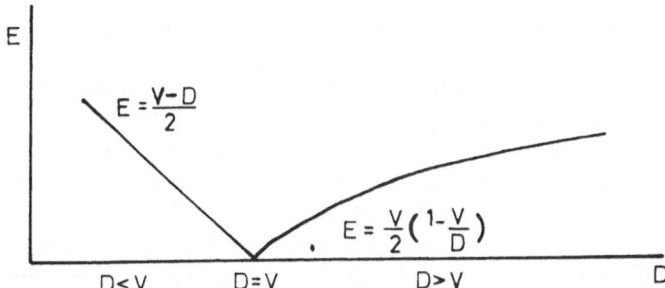

Fig. 4. The return E to members of a Hawks and Doves population as a function of D, the cost of serious injury. The return is minimum when $D = V$, the gain from victory.

Treisman noted that the average return to an individual in such a population increases as the danger of the escalated combat increases. The more dangerous a type of Hawk that has arisen, the greater the overall payoff. He clearly expected this result might be applicable to real animals, since he wrote:

> The existence of a few members of the species who are potentially highly destructive results in an overall population strategy which increases the expected return for all . . . Animals . . . possessing a choice and which can discriminate will do better to join heterogeneous (i.e., where $D > V$) rather than homogeneous populations (i.e., where $D \leq V$) assuming they adopt the appropriate behavioral strategy in each case. (1977, p. 201).

It is a mistake to extend the predictions of the model in this way. Although the model incorporates three mechanisms for settling disputes, only one of them involves an explicit cost which may represent what occurs in nature. Disputes between Doves are settled by a mechanism which carries no cost and is not discussed in detail. Consider a population in which $D \gg V$; in Treisman's terms, one that is "policed" by a few potentially highly destructive animals. In this population, most disputes are between Doves, and by assumption carry no cost. When a large proportion of the disputes are decided without any costs in this way, it is hardly surprising that the average return is greater than when $D = V$, for only at this point is the cost of disputes fully accounted for by the assumptions about cost that are made in the model. (At the point where $V = D$, all the population are Hawks, and the assumption about the cost of injury is sufficient by itself to explain why an injured animal stops fighting: If the injured animal persisted, its chance of sustaining the next injury would be 1/2, and so the expected cost of another round, $D/2$, would be equal to $V/2$, the expected return from victory. There would be no incentive for it to continue to fight, and whatever the effect of the injury on its fighting ability, its occurrence could decide the contest.)

We have suggested that one factor leading to Treisman's result might be the neglect of the cost of display in the standard Hawks and Doves model. What happens when we take this cost into account? Maynard Smith (1976) has provided a formula which includes the cost of threat (see caption to Fig. 3). In this case, we should expect the cost of threat to be $V/2$, by analogy with the War of Attrition. Substituting this value we find that the expected return to members of the population is independent of the proportion of Hawks (i.e., of the value of D) when $D > V$.

Similar arguments can be applied when $D < V$. For instance, when D is very small, why should an individual stop fighting for a valuable prize as the result of a very slight injury?

Thus, Treisman's predicion is one which depends on a pair of decision rules (allowing decisions in contests between Doves, and between Hawks when $D < V$) which are unrealistic, because they are not explicitly justified on biological grounds, and it is a mistake to extend it to real contest situations. But the example raises two further points of importance:

First, there is a relationship between the models and they are not really alternatives. We can imagine changing the value of V gradually and moving from a population where almost all the contests were Wars of Attrition using display, through a Hawk/Dove mixture, to a situation where all contests were escalated, and a War of Attrition might again occur. [At least one recent paper has shown that if the opponent does not give up when the interaction has escalated to the maximum level possible, the contest continues indefinitely, and Hyatt and Salmon (1978) explicitly apply the term War of Attrition to this terminal behavior.]

It would be surprising if there were marked changes in the average payoff as the value of V was changed in this way, and we may expect the value of $V/2$ for the cost of disputes to be a basic feature of some importance, at least in models in which the contestants are in the same internal state, so that the reward is of equal value to both. [Bishop *et al.* (1978) have shown that there is a lower cost when the value of the reward differs for the two contestants.]

Secondly, do the Doves benefit from being dove-like as opposed to escalating? In the original model, Doves were assumed not to escalate because of the cost of escalation. But what if there is also a cost involved in threat, and if it may occasionally be very high (as it will be under the War of Attrition)?

It could be argued that display only rarely involves a high cost, whereas this is required in every escalated contest. But suppose a new gambit arose which involved escalating to the same level as Hawks, but withdrawing after a suitable period of time even if no injury had occurred. The principle involved could be the same as that which allows animals to decide when to terminate display, and the period could be adjusted so that the occasional serious injury produced an average cost of $V/2$. I shall call this gambit the "Prudent Hawk."

Figure 5 shows the payoff matrix for this game. To give the Doves a chance, I have reverted to the convention that threat carries no cost. The new gambit would always win over Doves; it would also sometimes win over Hawks, although it would sometimes be injured in these contests. When the probability of injury in contests between two Prudent Hawks is V/D (so that the cost of these contests is equal to their average payoff, $V/2$), the ratios of types are:

$$\text{Hawks} \qquad \text{Prudent Hawks} \qquad \text{Doves}$$
$$1/(\alpha-1) \quad : \qquad\quad 1 \qquad\quad : \quad 1/\alpha$$

where $\alpha = D/V$, the "riskiness" of escalating. When $\alpha = 2$, Hawks and Prudent Hawks each form 40% of the population, but when $\alpha = 8$, Prudent Hawks have risen to 79%. Thus by escalating, but stopping when prudent, an individual can do very well, and under this model most combats should be escalated.

However, few should lead to serious injury — most would stop before this occurs. Intuitively, this seems to fit the biological facts better than the

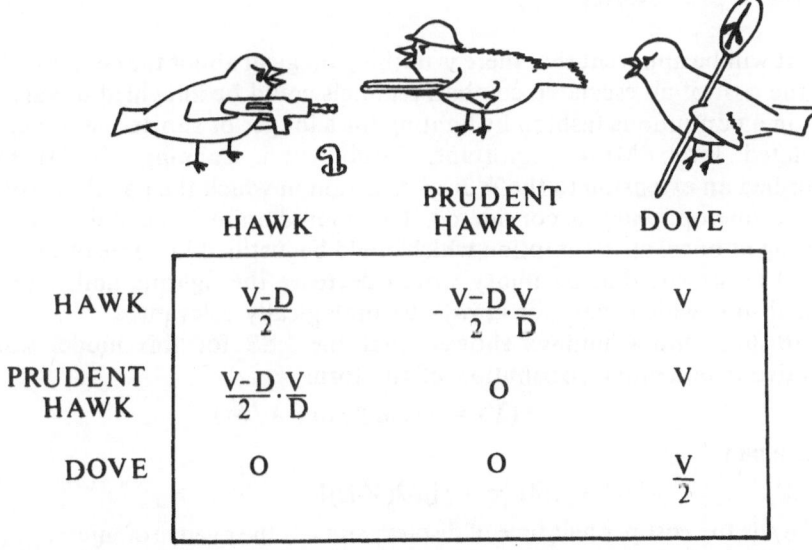

	HAWK	PRUDENT HAWK	DOVE
HAWK	$\dfrac{V-D}{2}$	$-\dfrac{V-D}{2}\cdot\dfrac{V}{D}$	V
PRUDENT HAWK	$\dfrac{V-D}{2}\cdot\dfrac{V}{D}$	0	V
DOVE	0	0	$\dfrac{V}{2}$

Fig. 5. The Prudent Hawks game. Payoffs and Symbols are as for Fig. 3. Note that Prudent Hawks always win over Doves, and that in interactions between Hawks and Prudent Hawks, the payoffs are equal and opposite.

original model in which all escalated contests ended in serious injury. Geist's (1971) review of data on Moose and other species showed that it involved risks of injury or death (respectively) of about 10% and 4% per year, not per contest, while data for mule deer give an estimate of 10% per year as the chance of injury (Geist, 1974) and data for musk oxen give values of 5% to 10% per year for the chance of death (Wilkinson and Shank, 1977). Clutton-Brock *et al.* (1979) have commented on the fact that contests between red deer stags rarely progress to the point at which serious injury occurs.

To return to my doubts about the interpretation of Hawks and Doves; I had asked whether models of this type might explain the variation in the physical level of escalation of contests. But on closer analysis, the crucial aspect of the model seems to be not the risk of serious injury, but the certainty that this will occur to one or a pair of rival Hawks. The model is relevant to the question "Why do animals rarely fight to the death?," which Maynard Smith and Price (1973) did in fact ask in their early paper, but which they did not distinguish clearly from a separate question, of greater interest to ethologists, about the occurrence of ritualized combat and threat. The model is not relevant to the latter question, and gives little help in explaining the biological phenomenon of escalation.

G. The War of Nerves

It will be apparent that there is nothing magical about the choice of $V/2$ for the cost of an escalated combat. Animals could be imagined to vary the cost in a continuous fashion by fighting for a longer or shorter period at the escalated level. More important, Bishop and Cannings (1978) have described an extension to the War of Attrition in which the possibility of injury to one opponent is considered. The injured animal is assumed to stop fighting at once, an assumption which could be justified in terms of Parker's (1974) argument that an injury would decrease the fighting ability of the animal, but which may not always be biologically relevant.

Bishop and Cannings showed that the ESS for this model was a negative exponential distribution of the form

$$P(X) = (1/M) \exp(-X/M)$$

with mean

$$M = V/[\mu - \lambda(V-D)]$$

where μ is the cost per unit time of display, and λ is the chance of injury.

Although Bishop and Cannings place little emphasis on the model, it seems important enough to warrant a name of its own, and so I have called it the War of Nerves.

Clutton-Brock *et al.* (1979) have described the fights of red deer stags as Wars of Attrition in which the rate of payment of costs mounts with the duration of the contest, with the additional possibility that a false move may lead to injury to either contestant at any stage after the two animals have locked antlers. This description is that of a War of Nerves in which the cost per unit time, μ is a function of time, $\mu(t)$, rather than a constant.

H. Is Risk Important?

I have concluded that discrete models of the type so far discussed give little help in understanding escalation, in the ethological sense of the term. What other factors may determine the choice of a level of escalation?

Consideration of the War of Nerves (or of the Prudent Hawk model discussed earlier) shows that the greater the severity of the injury that might be suffered, the greater the likelihood that the contest would end without injury to either opponent. Most contests would involve a very low cost (that of a brief interaction without injury) while a few (where injury occurred) would involve a very high cost. The variance of the payoff over a series of contests should thus be greater than for those fought at a level of escalation at which injuries were either less likely or less severe, and the dominant component was the cost of the display or other type of noninjurious interaction.

I found it easier to confirm this point by simulation than to approach the problems mathematically. In the elaboration of the Hawks and Doves model in which Doves settle contests by means of a War of Attrition, the Hawks were found to have a much greater variance of payoff than the Doves, over a series of encounters. In the War of Nerves, the variance was greater the greater the size of the term reflecting the cost of injury relative to that reflecting the shared cost of combat without injury. Where the chance of injury was zero, the variance of payoff is independent of the cost per unit time of the display or other form of combat.

In these models, the mean payoffs to individuals adopting different tactics are the same, but the variance differs. The greater the variance, the greater the chance of doing very badly over a series of contests. Could this be a factor which was important in determining the choice of level of escalation? Do animals adopt nonescalated tactics to reduce the riskiness of combat?

Theorists have assumed that it is the mean payoff over a series of encounters which is important, and have not discussed the variance. This is equivalent to the assumption of infinite credit. If we were considering, for instance, the gains and losses incurred in contests over food in a small Passerine this would imply that all the losses in December, no matter how severe, could be compensated by a spell of good luck in January. In many real situations it would presumably not be possible to compensate in this way; a Passerine that did very badly over the course of a single December day might perish during the following night.

The assumption of "limited credit" leads us to ask how an individual should choose its tactics over a series of contests to avoid "going bankrupt." If credit was particularly limited for certain categories of individual, e.g., the young or sick, we might expect them to chose a Dove-like, low-variance strategy. In species in which an unusually high success in contests could be reflected in an equivalently large number of offspring sired by the successful contestant we might expect escalation to occur more readily than it would in those species in which there was a lower variance in the number of offspring per contestant.

It should be clear from the preceding paragraphs that arguments such as these presuppose that success is measured in some intermediate currency, such as energy, and that the balance between profit and loss in this currency is mapped onto the ultimate scale of fitness. Previously, games theorists have dealt directly with measures of fitness, and these should (by definition) eliminate problems such as these; for instance, the cost (in units of ultimate fitness) of a given type of injury or period of display would not be constant, but would depend on the time and circumstances in which it was incurred. But it is probably realistic to look at the primary effects of most fighting in terms of a currency such as energy; for most of the reproductive cycle, the chance of reproducing immediately is zero, and the contribution of behavior to the probability of reproduction must take the form of an investment in energy, status, etc., which will later affect the chance of reproduction. The importance of the riskiness of particular tactics has also been noted with reference to the choice of flocking or solitary feeding strategies (Thompson et al., 1974).

I. Contests with Small Injuries

In the War of Nerves, an important assumption is that the injury (which is the only component of cost that is not partitioned equally between the opponents) is so serious that the injured animal stops fighting.

The increments in cost will not always be equally partitioned; in contests in which many of the injuries are slight and do not reduce the individual's fighting abilities (e.g., *Betta* contests, perhaps), equally matched opponents may receive injuries in a random sequence, each involving an increment of cost to one opponent. Alternatively, rather than receiving an injury, one opponent might choose to perform a more energetic display, involving a greater cost than its rival's less energetic behavior. How can continuous models be extended to cope with this type of contest? In this section I can do no more than sketch the outlines of a possible answer to this question.

It is useful to think of the cumulative cost of the fight to each opponent as being plotted along separate axes (Fig. 6). If the axes are at right angles, and if the costs are equally partitioned and the opponents equally matched, the progress of the contest will be represented by movement along the 45° line (line a in Fig. 6). (If the cost function was nonlinear, the speed of movement along this line would not be constant; for instance, it would increase with distance from the origin if the cost function was increasing.) In the right-hand panel of Fig. 6, the probability density functions $P(X) = (1/V) \exp(-X/V)$ representing the probability that each opponent will give up (under the assumptions of the War of Attrition) have been plotted separately along the two axes. The function along axis 1 represents both the probability that opponent 1 will retreat and also (because this will give victory to opponent 2) the reward function to opponent 2.

If opponents differed in ability (for instance, if one was smaller and had to exert itself more to match its rival's display) the asymmetry would lead to movement along a line away from the 45° axis (e.g., line b). Once the asymmetry became apparent it would be worthwhile for the weaker animal to give up. Even in a contest in which it was equally matched, the expected cost would be $V/2$, i.e., on average to win it would have to invest as much as it could expect to gain from winning or would require to find an equivalent

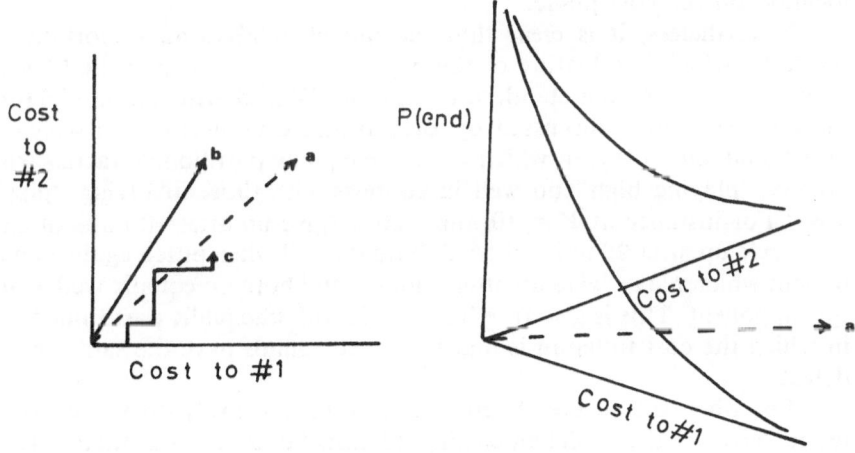

Fig. 6. See text for explanation.

to the contested item (Maynard Smith, 1974). When moving along line b, it must invest more than this to have the same probability of winning (i.e., to move a given distance to the right in Fig. 6). In this situation, it would require a lower investment to look for an alternative item. Conversely, it would be worthwhile for the stronger animal to carry on fighting when the asymmetry became apparent.

Line c in Fig. 6 shows the course of a contest in which the cost involves a series of small injuries, and is therefore unequally partitioned. If the opponents are equally matched, the overall gradient of the trajectory will be 45°, but by chance the line may depart considerably from the 45° line in some places.

Are the principles governing an animal's choice of tactics in this situation similar to those in situations in which the cost is equally partitioned? The only theoretical analysis which appears to be relevant here is Maynard Smith and Parker's (1976) discussion of a contest involving rounds, each ending with an injury to one opponent, which might or might not give up at this point. They examined tactics which involved giving up if injured in the first, second, etc., round (and also analyzed the effects of asymmetry in fighting ability; I shall only be concerned here with their results for symmetrical contests).

It is difficult to generalize from the authors' results, since they used a cost function for the successive rounds which increased exponentially, and thus much more steeply than those usually considered for a War of Attrition, and also because they assumed that the decision of whether or not to stop fighting was based on the result of the immediately preceding round alone, and independent of the earlier rounds, which would also affect the location on the cost plane.

Nevertheless, it is clear that the model involves an important new property, which is a feature of the payoff matrix rather than the ESS that may arise from it. In a standard continuous War of Attrition model (or in the discrete models discussed by Bishop and Cannings (1978) which are direct analogues of it) in which costs are equally partitioned, tactics which involve "playing high" do well in contests with those involving "playing low." For instance, if $V = 10$, the tactics "give up after 10 units of cost" and "give up after 20 units of cost" both do well when pitted against an opponent which plays "give up after 1 unit," and both do equally well against this opponent. This is also true for models with unequally partitioned costs in which the cost function is linear (i.e., all rounds have the same cost for defeat).

But when the successive rounds have progressively greater costs, the tactic "give up after a defeat costing 20 units" does less well than "give up after a defeat costing 10 units," even when both are pitted against opponents playing "give up after 1 unit," because the individual playing the more hawkish tactics may suffer by chance a long run of defeats, even against a very dove-like opponent. This disadvantage to individuals making plays which are greater than V, the value of the reward, is in addition to the disadvantage to such individuals of the losses in combat against others like themselves, and this increases the pressure which tends to eliminate from the population tactics which involve playing greater than V.

The effect is very clear when the cost function is steep, and for symmetrical contests, Maynard Smith and Parker (1976) found a bimodal ESS which contained only two tactics. I have confirmed this result in my own simulations using their very steep cost function; but when a more gradual cost function is used, the dynamics become much more complex, and it is still not clear whether one or more true ESSs exist, and from what range of initial populations they could evolve.

J. Contests as Random Walks

Suppose that instead of defining the costs of each round, we were to relax the rules and allow the contestants to chose the size of the next step at each point. Is there a rule governing the size of step that should be chosen at a particular stage in the contest?

Simpson (1966) drew an analogy between Fighting Fish and bidders at an auction to explain the pattern of escalation and matching seen in the contests. Here, I shall speculate that it might be fruitful to consider the contest as a random walk in the plane defined by the two cost axes in Fig. 6, and to look for analogies with the rules for successful gambling. It is striking that in real contests it is not necessarily the winner-to-be that escalates. Thus in the Fighting Fish, the eventual loser is as likely as the eventual winner to bite first (Simpson, 1966). Does this readiness to escalate in the animal which (at least on the conventional view) is ultimately proved the weaker reveal a parcllel to the gambler's best strategy in a game biased against him — if it is worth continuing, play boldly (Feller, 1968)? Or is the conventional view incorrect? Perhaps, instead of providing a "true" reflection of relative endurance, the results of such contests (in which the opponents are not grossly mismatched) merely demonstrate the effects of chance in determining which of the two opponents, each playing a distribution similar to those found in the War of Attrition, decides to give up first.

The reader will note that my argument that the weaker animal should play boldly if it is worthwhile for it to continue fighting presupposes some asymmetry in payoff, which compensates for the difference in fighting ability. This suggestion has received some support from the analysis of rutting behavior in red deer, in which it seems likely that solitary stags are in general inferior to harem-holders in fighting ability (Clutton-Brock *et al.*, 1979). Clutton-Brock *et al.* noted that solitary stags which initiated combat with harem-holders stood to gain hinds, but could obviously not lose any, while the converse was true of holders which initiated fights with solitaries (since their hinds could be stolen by kleptogamists). They suggested that this difference in payoff could explain the observation that solitary stags initiated fights with holders more frequently than expected by chance, and holders initiated fights with solitaries less frequently than expected, but they did not discuss the relationship of events within the fight to this asymmetry in payoff.

V. DISCUSSION

The basic Hawks and Doves model, on which most of my argument has been concentrated, is one which few, if any, games theorists would attempt to apply to most real contest situations. It does not, for instance, deal with the asymmetries which are present in many contexts and which can be shown to have an importance in games theory (Maynard Smith and Parker, 1976). But because it is so simple, it does illustrate particularly clearly the nature of the assumptions on which its predictions depend, and similar assumptions are important in more complex models.

One important conclusion, stemming from my analysis of this model, is that games theorists should discuss more explicitly the reasoning behind their choice of assumptions [e.g., Maynard Smith and Price's (1973) assumption that the contribution of wasted time to the payoff of the contests they modeled would be low relative to V, the value of the reward; Treisman's (1977) assumption that the "Super-Hawks" he discussed could defeat ordinary Hawks] and the relationship between the different theoretical models.

Ethologists have recently begun the task of bridging the gap between games theory and ethology (e.g., Clutton-Brock *et al.*, 1979; Davies, 1978). In view of this it might seem strange, in a book devoted to breaking down the barriers between different approaches to behavior, to find a chapter which urges the need for caution in applying models from games theory to real contest behavior. To understand why this emphasis is not incompatible with my earlier claim that games theory can make a real contribution to ethology, it is necessary to appreciate the nature of games-theoretic models.

Whatever their independent life in mathematics, in their application to behavior these models are *"minimal models"* [in Ludlow's (1976) terminology], models which have a status similar to that of a null hypothesis. When a model does not provide a full and realistic description of the system being modeled, we should not attempt at first to fill in its details, but rather play it against models from other categories, looking in particular for ways in which one type of model can be invalidated with respect to certain ethological contexts.

Thus, in the present chapter I argued that ethological accounts of escalation in various species described a progressive change in the level of physical interaction between two opponents. But while discrete models also postulate different levels of escalation, detailed analysis of the Hawks and Doves model showed that it was not relevant to the problems discussed by ethologists. The arguments suggested that the continuous models might in principle be extended in a way that would allow them to cope with these progressive changes.

An alternative approach to testing and possibly invalidating models is to pay special attention to models giving rise to predictions which run counter to generally accepted views; such models can either be readily disproved by comparison with ethological evidence, or can make the most conspicuous contribution to ethological thought. Dawkins and Dawkins (1974) say that such models, where they are successful, have a high Predictive Information Value. In my earlier paper (Caryl, 1979) I was able to show that there was a surprising degree of support in existing data for the prediction that agonistic displays should *not* transmit information about the probability of attack, a prediction which is directly contrary to the general ethological view (Caryl, 1979). The reanalysis raised several important new points; for instance it illustrated the need to identify which dimension(s) of information in a display is used by the receiver, and showed the lack of consistency (both from study to study, and from week to week) in the predictions about future actions that can be made from aggressive displays. These points suggest directions for further strictly ethological analysis. Thus, games theory can make a direct contribution to ethological work. The arguments in my earlier paper complement those set out here, but it would be inappropriate to repeat them in detail in this chapter, and the reader should consult the earlier paper in conjunction with the present one.

An implication of my argument about the nature of games-theoretic models, and the material discussed in the present paper, is that as yet we know very little about the properties of the models that are relevant to real ethological questions. At this stage, an understanding of the value and applicability of such models is more likely to come from the pragmatic approach of developing and attempting to apply new models, rather than from asking general questions about whether the models should be applied at all (cf. Oberdieck and Johnson, 1978).

A similar point about the value of looking for hypotheses which can be pitted against one another can be made with reference to ethological work. Not all ethological ideas about agonistic communication have the merit that they are testable. For example, Stout's (1975) definition of the "threat" content of a gull display in terms of its effectiveness in causing the rival to flee means that it is difficult to discuss the selective advantage of responding appropriately to these displays (as is required if we are to consider the evolution of the communication system) without circular argument. (Stout argued explicitly that the threat content need *not* be correlated with the overt probability of attack after the display; if the display alone caused the opponent to flee, the change in the stimulus situation would remove the need for overt attack). Such interactional explanations are not easy to test, and make it difficult to propose any coherent account of the forces leading to the evolution of displays with these properties. But it is relevant that when confronted with a "sticky" opponent (one that will not retreat — or a model) the gulls tend to persist in displaying, or use other "waiting" behavior, rather than attacking (Amlaner and Stout, 1978). This response to the refusal to flee is not the one that would be expected in terms of the interactional hypothesis mentioned, and it is plausible that one important factor in these contests is the time spent in the display.

One further general point about the ethological approach to communication may be worth noting: The common assumption is that agonistic interactions must involve communication, and that if an observing ethologist can see differences between the behavior of the winner-to-be and the loser-to-be, these differences will be used by the animals in communicating. But communication is a term which is appropriate when the sender's and receiver's interests coincide, and it is perhaps misleading to apply it also to a situation in which there is a conflict of interests. Like the ethologist, a Martian observer might well be able to detect differences in the tactics used in human wars which could be statistically related to their outcome, but we should not think it appropriate to describe these as communicating between the opposing sides. It is perhaps surprising that in choosing a word to describe the process underlying aggressive interactions between animals, we should select a term which has a root meaning *sharing*, and not one that describes human strife.

VI. ACKNOWLEDGMENTS

I am grateful to David Flitton and Tom Pitcairn for their comments on an earlier draft of this manuscript.

VII. REFERENCES

Amlaner, C. J., and Stout, J. F. (1978). Aggressive communication by *Larus glaucescens*. Part VI: Interactions of territory residents with a remotely controlled, locomotory model. *Behavior* **66**:223–251.

Bishop, D. T., and Cannings, C. (1978). A generalized War of Attrition. *J. Theor. Biol.* **70**:85–124.

Bishop, D. T., Cannings, C., and Maynard Smith, J. (1978). The War of Attrition with random rewards. *J. Theor. Biol.* **75**:377–388.

Caryl, P. G. (1979). Communication by agonistic displays: what can games theory contribute to ethology? *Behavior* **67**:136–169.

Clutton-Brock, T. H., Albon, S. D., Gibson, R. M., and Guiness, F. E. (1979). The logical stag: adaptive aspects of fighting in Red deer (*Cervus elaphus* L.). *Anim. Behav.* **27**:211–225.

Davies, N. S. (1978). Territorial defense in the Speckled Wood butterfly (*Pararge aegeria*): the resident always wins. *Anim. Behav.* **26**:160–173.

Dawkins, M. and Dawkins, R. (1974). Some descriptive and explanatory stochastic models of decision making. In: McFarland, D. J. (ed.), *Motivational Control Systems Analysis*, Academic Press, London, 323 pp.

Dawkins, R., and Krebs, J. R. (1978). Animal signals: information or manipulation? In: Krebs, J. R., and Davies, N. B. (eds.), *Behavioral Ecology: An Evolutionary Approach*, Blackwell Scientific Publications, Oxford, 494 pp.

Dow, M., Ewing, A. W., and Sutherland, I. (1976). Studies on the behavior of cyprinodont fish. III. The temporal patterning of aggression in *Aphyosemion striatum* (Boulenger). *Behavior* **59**:252–268.

Feller, W. (1968). *An Introduction to Probability Theory and Its Applications*, Vol. 1 (3rd ed.), John Wiley, New York, 509 pp.

Geist, V. (1971). *Mountain Sheep. A Study in Behavior and Evolution*. University of Chicago Press, Chicago, 383 pp.

Geist, V. (1974). On fighting strategies in animal combat. *Nature* **250**:354.

Harrison Matthews, L. (1964). Overt fighting in mammals. In: Carthy, J. D. and Ebling, F. J. (eds.), *The Natural History of Aggression*. Institute of Biology Symposium No. 13, Academic Press. London. 159 pp.

Hyatt, G. W., and Salmon, N. (1978). Combat in the fiddler crabs *Uca pugilator* and *U. pugnax*: a quantitative analysis. *Behavior* **65**:182–211.

Laudien, H. (1965). Untersuchungen über das kampfverhalten von Männchen von *Betta splendens* Regan (Anabantidae, Pisces). *Z. Wiss. Zool.* **172**:134–178

Ludlow, A. R. (1976). The behavior of a model animal. *Behavior* **58**:131–172.

Maynard Smith, J. (1974). The theory of games and the evolution of animal conflicts. *J. Theor. Biol.* **47**:209–221.

Maynard Smith, J. (1976). Evolution and the theory of games. *Am. Sci.* **64**:41–45.

Maynard Smith, J. (1977). Parental investment: a prospective analysis. *Anim. Behav.* **25**:1–9.

Maynard Smith, J., and Parker, G. A. (1976). The logic of asymmetric contests. *Anim. Behav.* **24**:159–175.

Maynard Smith, J., and Price, G. R. (1973). The logic of animal conflict. *Nature* **246**:15–18.

McCleery, R. H. (1978). Optimal behavior sequences and decision making. In: Krebs, J. R., and Davies, N. B. (eds.), *Behavioral Ecology: An Evolutionary Approach*, Blackwell Scientific Publications, Oxford, 494 pp.

Neil, E. H. (1964). An analysis of color changes and social behavior of *Tilapia mossambica*. *Univ. Calif. Publ. Zool.* **75**:1–58.

Norman, R. F., Taylor, P. D., and Robertson, R. J. (1977). Stable equilibrium strategies and penalty functions in a game of attrition. *J. Theor. Biol.* **65**:571—578.

Oberdieck, F., and Johnson, C. W. (1978). An evaluation of Maynard Smith's models of sexual strategies. *Anim. Behav.* **26**:631—632.

Parker, G. A. (1974). Assessment strategy and the evolution of fighting behavior. *J. Theor. Biol.* **47**:223—243.

Rand, W. M., and Rand, A. S. (1976). Agonistic behavior in nesting Iguanas: a stochastic analysis of dispute settlement dominated by the minimization of energy cost. *Z. Tierpsychol.* **40**:279—299.

Riley, J. G. (1979). Evolutionary equilibrium strategies. *J. Theor. Biol.* **76**:109—123.

Simpson, M. J. A. (1966). An experimental analysis of the threat display of the Siamese Fighting Fish, *Betta splendens*. Unpublished Ph.D. thesis, Cambridge University, England.

Simpson, M. J. A. (1968). The display of the Siamese fighting fish, *Betta splendens*. *Anim. Behav. Monogr.* **1**:1—73.

Stout, J. F. (1975). Aggressive communication by *Larus glaucescens* III. Description of the displays related to territorial protection. *Behavior* **55**:181—208.

Thompson, W. A., Vertinsky, I., and Krebs, J. R. (1974). The survival value of flocking in birds: a simulation model. *J. Anim. Ecol.* **43**:785—820.

Treisman, M. (1977). The evolutionary restriction of aggression within a species: a game theory analysis. *J. Math. Psychology* **16**:167—203.

Wilkinson, P. F., and Shank, C. C. (1977). Rutting-fight mortality among musk oxen on Banks Island, Northwest Territories, Canada. *Anim. Behav.* **24**:756—758.

Wilson, E. O. (1975). *Sociobiology*, Belknap Press, Harvard, 697 pp.

Chapter 11

SCIENCE AND THE LAW: A MUDDLED INTERFACE

Kenneth R. Wing

School of Public Health
School of Law
University of North Carolina
Chapel Hill, North Carolina 27514

All disciplines, in fact all thinkers, encounter a similar problem: the difficulty of sorting out their perceptions of the world from the biases inherent in their theoretical orientation, preferred means of investigation, and previous experiences. Ethologists, it seems, deal with the problem by tackling it head-on: explicitly acknowledging this difficulty and deliberately carrying out their endeavors in a manner that at least partially justifies a claim to an "open-minded and clear-headed style." (See Preface.)

This essay suggests that the disciplined clear-headedness of the ethologist be applied to a somewhat different line of enquiry than behavioral research: defining the nature of what we call "law" and sorting out the relationship that each of us has to it. The law— meaning laws, the legal systems, and the doings of legal institutions— is frequently misperceived or "muddled" by those in the scientific community as well as the rest of the world. Moreover, many of us and scientists in particular maintain a healthy disdain for the law and for the profession that it spawns. Everyone agrees that something ought to be done about the injustices of the law, but handicapped by our misperceptions, we rarely manage to change anything at all.

A realistic and open-minded view of the law would raise some intriguing questions. Is the law best approached as the work product of an insulated profession? Or are lawyers only puppets guided by others? Is the legal system based on principles or is it merely the reflection of men trying to dominate one another? Does the law lack what scientific method has to offer? Or is the problem of injustice relatively independent of legal methodology and orientation?

This essay attempts to raise these and other questions in a context that highlights their significance to those in the scientific community.

I.

It was six men of Indostan
   To learning much inclined,
Who went to see the Elephant
   (Though all of them were blind),
That each by observation
   Might satisfy his mind.

II.

The First approached the Elephant,
   And happening to fall
Against his broad and sturdy side,
   At once began to bawl:
"God bless me! but the Elephant
   is very like a wall!"

III.

The Second, feeling of the tusk,
   Cried, "Ho! What have we here
So very round and smooth and sharp?
   To me 'tis mightly clear
This wonder of an Elephant
   Is very like a spear!"

IV.

The Third approached the animal
   And happening to take
The squirming trunk within his hands,
   Thus boldly up and spake:
"I see," quoth he, "the Elephant
   Is very like a snake!"

V.

The Fourth reached out his eager hand,
   And felt about the knee.
"What most this wonderous beast is like
   Is mightly plain," quoth he;
"'tis clear enough the Elephant
   Is very like a tree!"

VI.

the Fifth, who chanced to touch the ear,
   Said: "E'en the blindest man
Can tell what this resembles most;
   Deny the fact who can,
This marvel of an Elephant
   Is very like a fan!"

VII.

The Sixth no sooner had begun
   About the beast to grope,
Than, seizing on the swinging tail
   That fell within his scope,
"I see," quoth he, "the Elephant
   Is very like a rope!"

VIII.

And so these men of Indostan
   Disputed loud and long,
Each in his own opinion
   Exceeding stiff and strong,
Though each was partly in the right,
   And all were in the wrong!

MORAL

So oft in theologic wars,
   The disputants, I ween,
Rail on in utter ignorance
   Of what each other mean,
And probe about an Elephant
   Not one of them has seen!

John Godfrey Saxe, *The Blind Men and the Elephant* (Ca. 1850)

However time-worn it may be, I have always relied on the fable of the blind men encountering their first elephant to describe how most people relate to the concept of law and to their legal system. That is, each of us tends to develop his own view of the law, and often to behave accordingly, based largely on anecdotal experience or infrequent encounters with courts, legislatures, or administrative agencies; and rarely on an integrated — or, in the case of a large segment of the public, informed — understanding of what the legal system or the governing process is or, just as importantly, is not.

As a useful metaphor, the encounter with an invisible elephant suffers from only one serious flaw: to the extent that the encounter is with a unified whole that is easily understood but poorly perceived, the metaphor fails. Law, meaning the substantive rules, the procedural machinations, and the "doings" of legal institutions, is hardly a unified whole in either method, substance, or effect; whether defined in terms familiar to jurisprudential scholars or in the terms of popular understanding, what we call "law" refers to a number of related but diverse things, moving in many directions, with a variety of conflicting purposes. (Were I to rewrite Saxe's poem, I might have my protagonists wander into a herd of elephants.)

But with that one exception, I find the elephant story quite useful and that it illustrates very well the posture many of us adopt with regard to the law, the legal system, and the people and institutions of which it is comprised.

Suggest the topic of law to a physician and he will predictably respond in terms that focus almost entirely on malpractice litigation. From his perspective, the law is a theatrical, high-stakes poker game played by avaricious lawyers representing disgruntled patients, naive judges, and emotional juries, usually at the expense of innocent medical practitioners. Make the same suggestion to a businessman and the inevitable discussion will concern the red tape and bureaucratic barriers that government officials — an amorphous "they" — use to encumber the diligent efforts of the business community. But ask those same government officials to describe the law and they will tell you of a chaotic world where the trumpet of the invisible elephant punctuates conversation at odd and inconsistent intervals, and where words like due process and equal protection and the right to privacy are used as swords and shields in the never-ending battle for power in the bureaucracy.

While to some "the law" is a barrier to business as usual, to others it is a technician's game for "protecting the comfortable insulations of privilege," at the expense of something called justice. And to many others, it is simply dismissed as the work of lawyers, those unprincipled entrepreneurs, the rights and writs of passage to buying a house, writing a will, getting married; the unreadable fine print at the bottom of the insurance form or on the back of the theatre ticket; the legal mumbo-jumbo that is the progeny of long-since forgotten Latin phrases.

Lawyers are not immune to this myopic posturing. Quite to the contrary, the lawyer not only encourages the rest of the world to continue the practice, he participates in it himself. The practicing lawyer's view of the law is almost antithetical to that of the legal scholar. To the scholar, the law is his research and his concern is primarily a description of the principles of law, with somewhat less concern for their application to a given set of circumstances. To the practitioner that does exclusively trial work, the law is the courtroom and the courtroom an arena for verbal gladiators; to his classmate that specialized in tax and finances, the law is confined almost entirely to the lawyer's office space; it is a complicated exercise in paper-shuffling and jawboning where success means not pleasing a judge but pleasing a client, or convincing him that he is pleased. The lawyer-turned-lobbyist may never see a courtroom or argue with a judge; the law is the legislative process; its interpretation and enforcement are only incidental; and success is striking the correct balance of political forces at the most advantageous time.

Those in the legal profession rarely reflect on the meaning of their endeavors to the rest of the world or grasp for an understanding of the whole of their field. Both the practice and the study of law are almost exclusively processes of continuing specialization and each speciality has its own peculiar view of the law, of the legal processes, of lawyering.

A moment's reflection makes this understandable: there are few incentives for the profession to concern itself with the broader picture or with describing that picture for the rest of the world. Even within academics, the scholar is rewarded for specialized expertise and for research geared to the rest of the profession, not the befuddled public. A few scholars of jurisprudence devote themselves to the study of law *per se*, but their work is for their profession (if *for* anybody), not the public. A review of the legal literature would reveal a few highly philosophical macroscopic perspectives on law and, of late, a rash of "this is the law on..." books for popular consumption, but rarely an attempt to explain the law or the legal system in a useful, understandable, but sophisticated way.

The only apparent difference between the posture of the lawyer and that of the rest of the world is that while the rest of the world suffers from its lack of understanding, the lawyer suffers little from his distorted or narrow vision, and gains handsomely from that of the rest of us. For as law is poorly or narrowly perceived, it becomes more difficult to master without assistance, particularly if we perceive the truly invisible elephant, the one which is not there at all. The world must pay to ensure against the lawsuit, real or imagined; to buy a lawyer's time (actually many lawyers, each being a specialist) to hurdle what appear to be senseless barriers or to interpret the pig Latin or to champion, however successfully, a cause.

The result is a strange legal environment: everyone disdains lawyers/lawsuits/legal process, yet everyone has a grumbling but healthy respect for law. The world roars with approval at Shakespeare's "Let's kill all the lawyers," but three centuries later no one has. The legal profession is alive and thriving and will continue to be so, secure in a blanket of public misunderstanding — about "the law."

This misperception of law may be dismissed as inevitable or even inconsequential if the problem is viewed as another example of extraprofessional or interprofessional conflict. We have all heard similar complaints before from economists, engineers, physicians, even truckdrivers: other professions — or the rest of the world — do not understand us or they misperceive what we do; the research or methodology of your field does not take sufficient account of that of mine; what I view as an important subspecialty you view as a preoccupation or "turf-building," and so on. And these familiar complaints are usually well taken. We all have the tendency to behave like blind men meeting their first elephant when we stray outside our own discipline or field of interest.

I want to emphasize, however, that to view law as a field of study, an academic discipline, or even as a profession, is to focus on rather incidental aspects of law. The law that causes us the most concern is not the law of scholars or even the law as it appears to various practitioners, but the law meaning the reality of the law in practice, the "doings" of legal institutions, and the impact of law on our lives and work and security. The one characteristic of the elephant that we should all agree on is that it is there, seemingly more so now than ever before, in a very real way. Our misperception of it may be inevitable, and many of us prefer that it be inconsequential, but nonetheless, it is something that we all encounter, or are dragged into, or have to participate in — or risk the consequences of active nonparticipation.

It is this law, which for convenience I will call the law in practice, that I believe is encountered but poorly perceived; it is this law that can and should be realistically understood.

Having impugned my own profession and most of the rest of the world, it may be safe to finally address those who are the primary concern of this article: the members of the scientific community and specifically those interested in animal behavior. While I realize I may now be doing to the scientist that which I claim the rest of the world does to the law, I am still willing to assert that the scientist is vulnerable to the same affliction that I have outlined above. That is, the scientist maintains a view of law, of the legal system, and of the governing process that is largely personalized and reflects what has historically been his rather infrequent encounters with the law. Thus some scientists adopt the narrow view that the substantive rules of law and the legal processes are primarily barriers to their work and threats to values or areas of conduct which are closely guarded in the scientific community, e.g., autonomy, academic freedom, the pursuit of knowledge; and they frequently color this view with a heavy dose of disdain, usually directed toward the legal professionals who are perceived as responsible for "the law." Or the law is seen in a more active mode, a pernicious force invading the workplaces of science intent on wresting "scientific" decisions from their proper place. There is frequently a marked we/they element in the scientists' view of law, a tendency to detach themselves from the law and dismiss it as the separate discipline of the lawyer or the politician or the bureaucrat.

As with the views of the rest of the world, I cannot deny that the scientists's perception is not at least partially correct. The law can be an obstacle; it can be a threat to valued interests; and surely questions of importance to science, both public and private, are finding their way out of the halls of academia and into legal decisionmaking institutions.

And make no mistake, the method of law, be it that of the scholar or of the practitioner, is decidedly unscientific in many respects. My concern is not that the scientist's view of "the law" ignores reality, but that his view is narrowly focused, frequently biased by anecdotal experience, and perhaps most importantly, it misplaces the responsibility for the character of the law and legal decisionmaking. By bringing in an element of detachment, the scientist shifts the responsibility for legal decisions to someone else, a responsibility I think we all have to share, as I will discuss in more detail below.

I have always found it somewhat ironic that a profession that prides itself in its disciplined objectivity is susceptible to this rather subjective approach to law. I also find it ironic — and, under most circumstances, unfortunate — that this tendency persists, given the fact that the scientific community, perhaps as no other profession with the possible exception of medicine, now finds itself so often in confrontation with the beast. In many of his endeavors, the scientist finds that he is confronting not just his professional colleagues but more than infrequently lawyers, judges, legislators, bureaucrats; institutional review boards with lawyer members, legal counsel, and something called consumer or public participation are reviewing his research protocols. Administrative law judges (do you really know what *they* are?) are making decisions on the availability of sugar substitutes, junk food, pesticides. The introduction into interstate commerce (I can assure you that "introducing into interstate commerce" includes a far broader range of activities than most people imagine) of laetrile has been forbidden by the Commissioner of the FDA, but criminal prosecution of certain laetrile users is enjoined by some state judges (enjoined?). And there are a variety of state and federal statutes, many with implementing regulations and explanatory materials called guidelines and handbooks (the legal significance of which is far from clear) limiting access to certain kinds of records and information, and forbidding certain kinds of research-related activities, each with a different set of enforcement mechanisms.

The law appears to be there, perhaps more or less often than it really is, and it demands attention. Particularly for those who are either concerned with the broader social implications of many research-related questions or for those who are unavoidably drawn into controversy because of those implications, the law must be correctly perceived and reckoned with in an effective way.

What is an effective approach to the law? In particular, how should the scientific community view legal processes, the legal profession, and the law in practice?

I cannot offer a definitive answer to these questions or a complete descriptive restatement of the law and the legal system. Rather what I intend to do is draw attention to certain characteristics of the law and the legal system which are of particular importance to the problem of misperception, characteristics which must be understood as a basis for a more realistic understanding of the law in practice.

I should begin with one caveat: my insistence that the scientific community revise its perception of the law should not be interpreted to mean that I accept or endorse the legal status quo. Quite to the contrary, the history of legal decisionmaking and of the legal profession, viewed from anyone's perspective or evaluated in virtually any way, is notably unimpressive. But changes, even where appropriate, may be difficult or even impossible. While we may prefer it to be otherwise, we must in the short term approach the law for what it is. Moreover, change, even where clearly justified and possible, must be premised on a firm understanding of what the actual problems and their underlying causes are. My description of the legal system will be quite unflattering — in fact, it will be similar in many respects to that of those who dismiss the law as the work of avaricious lawyers or the crazy game of bureaucrats. Nonetheless, what these attributes of the legal system suggest to me, in terms of change and in terms of how scientist-participants should respond, will be quite different than what they suggest to those who would detach themselves from the law or salute the Shakespearian call to arms.

To begin with, the law and the legal system provide us with several formal decisionmaking processes. There are sets of rules, means for creating and applying them, and means for their enforcement. There are, of course, other ways to make decisions and other rules that affect our lives. In fact, the jurisdictional dispute that questions whether a given problem is appropriately settled within the legal system is a primary one in law — although it may well rely on a false dichotomy as will be discussed below.

Apparently all societies have laws and have something roughly akin to the American legal decisionmaking institutions. There is clearly enormous diversity from society to society, but within this diversity there are some common elements of relevance to the present enquiry.

First of all, what we call legal decisionmaking is the way that certain kinds of controversies are resolved. That is, certain kinds of decisions become legal controversies, move beyond the individual, his family or social or religious group, to representatives of a larger group that are vested with the power of governing, i.e., the ability to enforce — with what legal scholars like to call "legitimized force" — compliance with their decisions. Legal decisionmaking is, in effect, a substitute or alternative for decisions or agreement among people or institutions that are determined by suasion, reason, bargain, pride, or other influence, but not by "legitimized force."

Some decisions are inherently legal controversies. In other situations the appropriate role of the legal decisionmaker may be the heart of the controversy. The choice to initiate legal decisionmaking or to advocate that a decision is appropriate for a legal decisionmaking process depends upon a number of variables: social or political climate, individual predisposition, availability of resources, actual or psychological cost of initiating controversy. But however it is initiated, once a decision falls within the arrangements by which these "extra-group controversies" are resolved, it becomes a very different kind of decision and, most notably, it becomes one that is enforced by the sanctions only available to government institutions.

The result is that legal resolution of a controversy is likely to have rather concrete and binding consequences. You can ignore, in the strict sense of the term, the decision of your family or social group; you can dissent from the opinion of your professional peers. Arguments between scholars are rarely resolved and can be lost today only to be won tomorrow. There can be real and binding consequences as a result of these other controversies, but they are decidedly different from the consequences that attend legal controversy. If you lose an argument with a judge or legislator or even certain species of bureaucrat, the outcome can be quite binding and very real. A legal decision can be right or wrong, poorly reasoned or maliciously intended; a final legal decision can be accurately described as solving nothing; but somewhere on the back page or the bottom line will be the binding resolution of something: for better or worse or more of the same, social benefit programs will be carried out or not, activities will be prohibited or required, wealth or power will be distributed or the existing distribution forcefully maintained.

Thus from the point of view of those who are affected, the various legal decisionmaking processes can become less important as a pursuit of the truth or justice or as a collective effort to carry out agreed-upon principles of social equity, and more important as a battle to prevail. The actual consequences of an outcome become the focus of attention. Winning and losing become surrogates for being right or being wrong; and there is a strong temptation to evaluate the means for achieving those outcomes solely by the ends that they achieve. When all is said and something is about to be done, I want things done "my way."

But maybe that is not so bad as it sounds to be. After all, as many of my colleagues would tell us, it is the genius of the American legal system that it not only tolerates the human tendency to maximize self-interest, it thrives upon it. The confrontation between highly motivated advocates in the legislative and administrative arenas is exactly what assures the legislator or bureaucrat that he is fully informed. And is it not the violent battle of adversaries before judge and jury and the corresponding balanced division of labor that ensures that the truth will be critically examined but impartially decided?

Perhaps this is an accurate view. But that is the essence of our problem: what is an accurate view? What is really going on? How does it work? What are people "up to?" And if legal decisionmaking does not work well, but it does affect me, what can or should I do?

First, consider the legislative process. the fact that legislative decisionmaking is predominantly political, i.e., controlled by power not principles or even party platforms, is usually acknowledged but often overlooked. The public gasps in what must be mock astonishment as powerful lobbyists dictate to legislators with a casual disregard for either long-term or short-term public good. They seek to maximize the self-interest of those they represent. Those of us who view ourselves as outside the process analyze policy, measure social utility, weigh costs and benefits; yet we should know that our research or the merits of legislation do not control outcome; the politics of a proposal will determine its fate and the relative power of its various proponents are the critical determining factors. The legislator cannot afford to consider simply the merits of a proposal, but who will win and who will lose *and* how powerful those winners and losers are. He cannot even afford to serve blindly the needs of his own constituents at the expense of good policy — a philosophical position that some scholars would tell us leads to good policy in the long run. Political decisions are not based simply on voter preference but on that preference after it is shaped and molded by forces available, again, to the wealthy and powerful, i.e., those who can afford such manipulation. So long as election campaigns are financed as they currently are, it is hard to imagine it being any other way. And the deeper one peers into the "real politics" of Washington, D.C., or our state legislatures, the more clearly the message becomes: legislation can be figuratively, perhaps literally, bought and sold.

Even when it appears that a legislative body is responding to issues and not political or economic power there is a decidedly quixotic nature to the process. "Squeaky wheels" are attended to and horror stories command attention today only to be forgotten tomorrow, as a seemingly arbitrary and often emotional atmosphere pervades the legislative process. Note, however, that the atmosphere may be emotional but hardly by accident; and to label as arbitrary any legislative decision is to ignore the underlying causes for legislative concern. Nothing happens by true accident, but always by someone's design. Legislation may be piecemeal, overreactive, capricious, or ill-conceived in outward appearance, but most often such legislation should also be viewed as an unflattering tribute to those interest groups who, having a realistic understanding of the legislative process, have managed to play the game politically and, therefore, successfully.

When the stakes are real and binding, the outcome becomes the predominant concern, particularly to those who have the ability, through power or wealth or guise, to influence that outcome. The concern can be based on the most noble objective, crass self-interest, or evil intent, or it can be a muddled combination of all three.

Judicial decisionmaking is different in many respects from legislative decisionmaking, but it also has parallel characteristics. In particular, the participants have a high degree of concern for those real and binding consequences and whom they impact upon.

In theory, we are sometimes told, the judge with the help of a jury impartially decides the facts, determines his best reading of the law, and makes a decision. In fact we are seldom told, all judges struggle with the temptation to work backward: what do I (or my friends or party or social class) want? How can I read the law to justify that outcome? Which evidence should I weigh and which should I disregard? Some judges give in easily to this temptation. Others claim to scrupulously avoid it. Most, I would imagine, tread water in between. No one really knows — and the issue has rarely been closely or openly examined. Only recently, for example, has there begun to emerge a realistic and informed picture of how the U.S. Supreme Court chooses its cases and develops its opinions. That picture includes honest, hard-working justices trying to extract specific decisions from vague general principles and conflicting evidence and struggling to find the "law." It also includes human beings with political sentiments, personal ambitions, and frequently an opinion of outcome not based on legal principle, but one that will be justified in those terms.

Outcome determinism is also reflected in the behavior of lawyer and litigant. For the lawyer, of course, the object is not achieving a just resolution *or* his own preferred outcome. It is winning, which is what he is paid to do. It is the rare client who pays a fee for achieving a just result or prevailing only if the law is only truly on the client's side. Some people begin with the request for an objective opinion, but most people think they are right, whatever the law says. And when the hypothetical turns to marching orders, the instructions are usually much simpler: win.

How do lawyers win? Certainly the lawyer uses persuasion, insightful argument, and logic; and there is nothing quite so righteous as the advocate whose position is firmly supported by legal principle or justice (or, on that rare wonderful occasion, both). But the skilled advocate has other weapons in his arsenal: an appeal to the emotions: the creation of an illusion; or dilatory tactics that cost one's opponent more than he is willing to pay for an outcome. Indeed, trial litigation is much more accurately described in those terms than any others. Lawyers are hesitant to discuss it in any but the most confidential terms, but litigation is basically a game and a game that is largely determined by time and effort considerations. Is it worth x dollars to file a suit? Or $x + y$ if the other side contests? Is your opponent willing to negotiate if a trial will cost z? Or will he if your pretrial tactics will double the cost? Is time on your side? Can he wait two years for a decision? Or will a six-month delay be tantamount to a victory even if we eventually were to lose on the merits?

If you could carry out a study by any even roughly scientific method to evaluate the factors that influence outcome in a judicial decision (something which, parenthetically, no one in the legal profession, practitioner or scholar, *ever* does) I would expect to find that of the ten factors most responsible for outcome, ranked in order, neither the correctness of the law, nor the equities of the case (holding aside the problem of defining what they are) could possibly be more than third or fourth in importance.

I am convinced that greater weight would have to be given to such factors as the relative resources available to the parties for investigation, preparation, and trial; the relative ability of the litigants to tolerate the status quo pending final outcome; and the relative competence, experience, and preparation of legal counsel.

Many of these same things can be said of the various executive and administration decisionmaking processes that make up the all-important yet often confusing third branch of government. While the rulemaking processes of state and federal government do not exactly parallel legislative decisionmaking, many of the same influential elements would have to be included in a realistic description of this aspect of the governing process. The direct and indirect pressures of the lobbyist, the concern for the political implications of decisions, the need to respond to public opinion, properly reasoned or not, all weigh heavily on administrative discretion.

The same is true of the quasi-judicial decisions of regulatory agencies and funding programs; the administrative officials who must interpret their legal authority and make fact-finding decisions are subject to the same temptations as their judicial counterparts. Lawyers as administrative advocates and those they represent have the same incentives to prevail and they behave according to the same calculated estimates as they do in formal litigation. I would expect that a similar study of administrative decisions designed to parallel that of judicial decisions described above would produce almost precisely the same results.

All this is not to deny the existence of the well-intentioned legislator or the disciplined and honest judge, or the capable and fair bureaucrat; all exist. Nor am I suggesting that all legal decisionmaking is hopelessly biased by rampant self-interest and the influence of the wealthy and powerful. My intention is not to condemn so much as it is to criticize. Just as I reject the views of those who oversimply or narrowly perceive the law or who would dismiss law (or lawyers) out of hand, I reject as well the views of those who would make it more than it is or adorn it with attributes that are really quite rare. In fact, I find it useful to reject any view of the law in practice that starts with the premise that it is, *in fact*, a system of justice, or a codified value system, or reflective of moral doctrine. My description of the legal system is intended to emphasize that the law, the legal decisionmaking processes, and the legal profession, are all part of the way in which certain kinds of controversies are actually resolved by human beings, no more and no less. Perhaps I have understated the influence of notions of justice or of a higher order. Or maybe there is a kind of "invisible hand" principle that guides all this self-interest and outcome orientation to just results. But where is the evidence of these phenomena? I prefer a view that assumes that these institutionalized arrangements do what they do and enforce their decisions without any design but with a kind of momentum that masquerades as legitimacy.

I prefer a description that focuses on law as the existing means for resolving "extra-group" controversies and one that includes the realities inherent in conflict resolution enforced by government sanctions. My central purpose here is fairly straightforward. I want to convince my readers that however one characterizes the law or behaves with respect to it, his assessment of the law should be based on a realistic understanding of the actual role played by law in our lives and of the true behavior of courts and legislatures and lawyers and litigants. One must know what the law is just as one must know what it is not, and be well aware of just how elusive the nature of the elephant can be.

For example, I have emphasized that a realistic view of legal decision-making and particularly judicial decisionmaking is that it is primarily the means by which certain kinds of controversies are resolved. It is distinguishable from many other kinds of decisions notably in terms of the kinds of controversies involved, the sanctions by which legal decisions are enforced, and the binding impact that legal decisions have. Therefore, any suggestion that a particular decision should be one made by science and not by law, or by scientists and not judges, would have to mean just that, i.e., one is expressing a preference that the decision ought not to be one enforced by governmental sanctions, a preference that by the nature of the controversy may not be a viable option. Or to express the example in other terms, if one prefers that scientists make decisions that are binding, one is asking that scientists become judges, not vice-versa. And the preference for the scientist-turned-judge must also consider the likelihood that whoever assumes the judicial role, regardless of training or professional identity, will be subject to the same influences and temptations that affect others who have played that role.

It could well be that judges or legislators or lawyers in general would benefit from additional training in scientific method. That is a different proposition. But to analyze the problem of outcome determinism or subjective decisionmaking in a manner that implies that the solution is to have legal controversies resolved in the same manner as research questions or academic disputes, i.e., in the manner of many scientists, is not a comparison of like situations. It also assumes that the undesirable realities of legal decisionmaking are attributable simply to the methodology of legal decisionmakers; that assumption remains to be tested. The problem may be determining who should make decisions, but also how should they be made fairly under circumstances where the outcome will be real and binding and will be enforced by governmental sanction. A scientist with the powers of a judge or a legislator, or even a bureaucrat, is not in reality functioning as a scientist. If he is given the role of legislator or judge, he would be a legislator or judge; would he not behave like one? Do judges or legislators behave as they do because they are fools, or is such an assumption a misperception of underlying cause and effect? Again, the answers are not clear.

We have some evidence of how that scientist might behave. Many of the administrative agencies of the state and federal government, including those most often criticized as being bureaucratic or politically biased, were designed to allow experts, including scientific experts, the authority to perform quasi-judicial and legislative roles. Those bureaucrats who make decisions for the FDA or constantly burden the flow of research funds from the government with paperwork and regulatory strings are likely to be scientists by training, but administrative agents of the government by role.

A related issue involves the distinction between questions of fact and questions of, for want of a better term, subjective value, and the delineation of the scientist's role in determining each.

From the decision of a bureaucrat to issue regulations governing research funds to the decision of a judge in a child custody trial to the legislative drafting of occupational health and safety legislation, many legal controversies begin with questions of relatively objective fact but require for ultimate resolution a choice between alternatives that can only be based on value preference. The perceptual problem involves correctly identifying the decisions being made and, particularly, the proper roles to be played in the decisionmaking process. The unforunate tendency is to be overinclusive, to assume that the qualifications of the expert to make certain decisions justify the expert's role in final resolution or in value choice, i.e., in the name of expertise, one expresses what is actually the desire to acheive a preferred outcome.

A legislative or judicial decision must obviously begin with a firm factual base — at least within practical limits. And if that factual base requires more than common understanding, then there is obviously an appropriate role for the expert/scientist. In some legal controversies that role is absolutely critical. In fact, a case can be made for the proposition that in our complicated, technological, and frequently specialized society, the role of expertise in legal controversy resolution can only expand; the facts are simply increasingly difficult to adduce, collate, and synthesize.

The major qualification of this line of reasoning is that the argument should not be overextended to delegate to the expert those decisions, or these aspects of decisions, that are not matters of data analysis or that do not require sophisticated understanding, but that are matters of values and choice between competing alternatives based on values. Utility analysis, or whatever one calls the weighting of the relative value of choices, has to be treated differently than the evaluation of any one of the choices. The expert may be necessary to define and analyze the differences between alternatives and delineate discrete choices (although the argument that an issue requires specialized expertise to understand is frequently overstated). But the expert is not necessarily qualified to make the ultimate choice. In fact, there are a number of reasons, e.g., personal or professional involvement, that may disqualify the expert from any participation in the choice among options. Expertise qualifies one to inform the legal decisionmaker, not assume his role.

Value judgments may, of course, be obvious. The value of a good research protocol, for example, over a bad one leaves so little room for discretion that the problem of identifying the proper decisionmaker becomes virtually irrelevant. Choice is fixed by the definition of options in many circumstances. But suppose the choice is between government funding (or allowing) good research which incurs a slight risk of catastrophic results to the experimental subject and less valuable research that is relatively safe. The choice of decisionmaker becomes critical and the qualification to delineate the choices and properly inform the decisionmaker in no way suggests qualification to make that ultimate choice.

Whether describing or criticizing the legal system, the assessment must begin with a proper and realistic definition of the problem. Declaring it a matter that only experts can understand and demanding it therefore be delegated to their judgment offers a solution by incorrectly defining the problem.

The tendency to hold the legal profession responsible for the injustices of the legal system can be questioned in a like manner. Undoubtedly, both legislative and judicial processes can produce costly and inequitable results and both can be played like a game where the powerful dominate the weak. But is the greed of the legal profession the underlying cause, and the benefit that accrues to the wealthy and powerful interest an incidental side effect? Or is it the other way around? The answers to those questions are absolutely critical to a realistic assessment of the law.

Any number of similar questions can be tested against the realities of legal decisionmaking. As a final example, consider the view that legal decisionmaking is more often a product of a determined battle to prevail among interest groups and not a product of a collective effort to produce just results: Does this not question the often-heard analysis of social controversy in terms that define the underlying issues as "appropriately" within or without the legal system — at least to the extent that the analysis looks to practice and precedent as indicative or appropriate? Under the law as practiced, little distinction can be made between a question or controversy that historically has been settled by other then legal decisionmaking owing to societal notions of justice, and those which courts and legislatures have been prodded and persuaded not to address — for the purpose of actively maintaining the status quo. You may enjoy personal autonomy or decisionmaking control over your family, your property, or your business affairs; to you it is only just and fair that these decisions are "left" to your discretion. But particularly in regard to those personal interests and areas of conduct also valued by those of wealth and power, legal decisionmaking has been very active in maintaining those interests, albeit in the name of nonparticipation. Consider, for example, all of the interests that are labeled private property and how actively courts and legislatures maintain property as it is distributed using language which suggests they are taking a "hands-off" position. The dichotomy between controversies within and without the legal system can be useful in analysis and description, but it also can be a verbal game, constitutional rights notwithstanding, and one often favored by the "haves" in their efforts to fend off the "have-nots." Or to put it in terms relevant to many of the controversies now before the scientific community, what appears to be the invasion of legal decisionmaking into the lair of the scientist can also be described as the reversal of a historical tendency of legal decisionmakers to consciously defer to interests of which the scientific community has most often been a part. The growing feeling that science now confronts law more often than it did before may be better characterized as a tremor emanating from a shift in the relative abilities of various individuals and groups to participate in the games of law rather than as the result of a shift or slippage in the legal system itself.

My purpose herein is not to quibble with the learned scholars of my profession or to enter into the philosophical debates that try to characterize the nature of the law. My purpose is far more basic. I want to do little more than make the point that courts, law, lawyers, *et al.* are part of something we all encounter but do not see very clearly.

Yet we are apparently unaware of the problem. We often behave with the sure confidence that we understand what "it" is all about and as if "it" is a simple unitary whole. When what we see, or think we see, does not please us, we act as if it can easily be held apart from us as if someone else is responsible, those silly lawyers or politicians or someone of that ilk.

We claim justice is done when we win; we withdraw indignantly when we lose; we argue that government has no right — a word I have struggled not to use in this article — to intervene when we are the "haves," and we complain that "they" have unfairly fixed the distribution of wealth when we are the "have-nots."

Frankly, I cannot understand how anyone, and most particularly those in the scientific community, can afford the luxury of maintaining the views that most of us have. If such views have been successful in the past, it is only a reflection of the often ignored fact that some of us, and particularly, again, those in the scientific community, despite our protestations have done rather well under the legal status quo. While I remain skeptical of the quasi-paranoic accounts of law and lawyers attacking our sovereign territory on all fronts, I do find that modern society is increasingly one of interdependence and conflicting needs, elements that lead to the kind of controversy that frequently draws us and our issues into the legal arenas.

What, then, is a good view of law, one that might be useful to bridge interdisciplinary chasms?

I have tried with some brief examples to illustrate what might be regarded as the tentative beginnings. I admit that I am much more certain of that which I want to reject than of the specifics of the view which I would prefer one relied on. I also admit that I am pushing in two directions at once: trying to give the law and the legal system what credit is deserved, but also trying to ensure that they are seen as no more than they really are.

If I must adopt one definitional model, the law appears to me to be a game where decisions are made by rather pragmatic actors. Decisions are formulated and reasoned, but also influenced and manipulated. Justice is important to some who play, but power, wealth, and social standing are also familiar faces. In the case of litigation, even a decision by a well-intentioned, truly objective judge will be based on some curious psychological influences on his thinking, by the decisions of previous courts — including firm decisions not to decide, and by a long list of practical considerations that may even keep the truly objective judge from rendering a final decision.

In the case of legislation, the game of power and wealth may be more readily seen. The experienced legislator, unlike his judicial counterpart, boasts of his political skill, an intriguing euphemism, and the rest of us first admit and then deny the true meaning of the word politics.

In all the various rulemaking and quasi-judicial functions of the state and federal bureaucracies, these games are played out in endless variation.

You can criticize these games: cumbersome, inequitable, antiquated, costly. You can readily document that governing is not always serving the public good and that it often serves other interests. Even when legal decisionmaking is not biased it can still appear to be capricious.

But another reality factor must be considered: The law and the legal system are the way they are not merely because of happenstance or human silliness, or anything as simple as "they are doing it to us," but because some people, including "us," want it that way. The interesting pattern of winners and losers must be traced, including one's own place in that pattern, if one wants to understand the law and why it is what it is.

There are any number of ways in which justice could be promoted; many ways legal decisionmaking could be more objective, expedient, scientific. But all that presupposes that some sort of justice is our common goal. Is it "theirs?" Is it "ours?" And, consider the intriguing outcome of the hypothetical situation wherein you start playing the games of law more equitably, but I do not.

What to do about all this is not clear. Many of the suggestions that are frequently made outside of the legal community sound to be either naive or ill-suited to legal decisionmaking. Many more simply ignore the substantial resistance to any change, particularly systemic change in the legal system, that will come from those powerful interests who benefit so much from the way things are. Most of the suggestions that come from within the legal profession sound little better and frequently evoke an understandable skepticism in the listener.

To me the answer to the problems of the legal system must be traced back to the distribution of wealth and power. Thus the only specific answers I can offer usually focus on ways to ensure that power is shared and diluted. Public participation, procedural due process, an informed everybody, may appear to be the kind of lawyer's response of which one should be skeptical. Like the elephant story, these notions may be timeworn, but nonetheless they seem designed to deal directly with the real problems of legal decisionmaking. While they are difficult to achieve, they at least are achievable, however inadequately adapted in many existing situations.

So long as the bulk of the public is kept at arms length and remains uninformed, the problems of legal decisionmaking will continue to be conumbral; the absence of public participation creates the opportunity for wealth and power to influence decisions; the ill-informed presence of the public does little good and frequently makes the public the unwitting tool of someone's advocacy.

Again, the answer, or answers, will likely begin with some sort of redistribution of the kind of power which affects legal decisions, but it can only begin there. Obviously utopian notions of major shifts in real power are little more than pipe dreams; marginal shifts, however, are quite realistic; so much could be gained if all of us did nothing more than take a renewed, sober look at the legal institutions that affect our lives, and the influences upon them, and the pattern of winners and losers that they produce, and accepted the responsibility for participation in legal decisionmaking.

REFERENCES

Beutel, F. (1975). *Experimental Jurisprudence and the Scienstate*, Fred B. Rothman, Littleton, Conn.
Boffey, M. (1976). Science court: high officials back test of controversial concept, *Science* **194**: 167.
Coughlin, G. G. (1972). *Your Introduction to Law* (Rev.), Barnes & Nobel, New York.
Council for Science and Society. (1977). *Scholarly Freedom and Human Rights*, Barry Rose, Ltd. for British Institute of Human Rights, Chichester, Eng.
Curlin, J. W. (1974). Fostering understanding between science and the law, *Conn. Med.* **38**:683.
Frank, J. (1973). *Courts on Trial: Myth and Reality in American Justice* (3rd ed.), Princeton University Press, Princeton.
Greenberg, D. S. (1976). Trial run for a "science court", *N. Engl. J. Med.* **294**:505.
Lefcourt, R. (ed.). (1971). *Law Against the People: Essays to Demystify Law, Order and the Courts*, Random House, New York.

Martin, J. A. (1977). The proposed science court, *Mich. Law Rev.* **75**:1058.

Talbott, R. E. (1978). "Science court" a possible way to obtain scientific certainty for decisions based on scientific "Fact"?, *Environ. Law* **8**:827.

U.S. News & World Report Editors. (1973). *What Everyone Needs to Know About Law*, S & S Press, Austin, Texas.

Weinberg, P. "Science court" controversy: are our courts and agencies adequate to resolve new and complex scientific issues?, *Record* **33**:8.

INDEX